D0085302

The Barefoot Expert

Recent Titles in
Contributions to the Study of Computer Science

Reckoners: The Prehistory of the Digital Computer, from Relays to the Stored Program Concept, 1935–1945
Paul E. Ceruzzi

Machines and Intelligence: A Critique of Arguments against the Possibility of Artificial Intelligence
Stuart Goldkind

The Barefoot Expert

THE INTERFACE OF COMPUTERIZED KNOWLEDGE SYSTEMS AND INDIGENOUS KNOWLEDGE SYSTEMS

Doris M. Schoenhoff

Foreword by
WALTER J. ONG

CONTRIBUTIONS TO THE STUDY OF
COMPUTER SCIENCE, NUMBER 3

GREENWOOD PRESS
Westport, Connecticut • London

Library of Congress Cataloging-in-Publication Data

Schoenhoff, Doris M.
The barefoot expert : the interface of computerized knowledge systems and indigenous knowledge systems / Doris M. Schoenhoff ; foreword by Walter J. Ong.
p. cm. — (Contributions to the study of computer science, ISSN 0734-757X ; no. 3)
Includes bibliographical references and index.
ISBN 0-313-28821-6 (alk. paper)
1. Expert systems (Computer science) I. Title. II. Series.
QA76.76.E95S383 1993
006.3'3—dc20 92-40225

British Library Cataloguing in Publication Data is available.

Library of Congress Catalog Card Number: 92-40225
ISBN: 0-313-28821-6
ISSN: 0734-757X

First published in 1993

Greenwood Press, 88 Post Road West, Westport, CT 06881
An imprint of Greenwood Publishing Group, Inc.

Printed in the United States of America

The paper used in this book complies with the Permanent Paper Standard issued by the National Information Standards Organization (Z39.48–1984).

10 9 8 7 6 5 4 3 2

For
Charles F. McDermott, SJ
(27 October 1923–26 April 1991)
with love and laughter

That is my present. . . . In one of the stars I shall be living. In one of them I shall be laughing. And so it will be as if all the stars were laughing, when you look at the sky at night. You — only you — will have stars that can laugh.

— Antoine de Saint-Exupéry, *The Little Prince*

Contents

Foreword

How does computerized knowledge, particularly in its advanced "expert system" forms, relate to other kinds of knowledge today? This question is particularly acute in the hundreds of cultures where computerization, and even management of knowledge by writing, is a rarity, but where knowledge is often in fact otherwise managed very effectively to meet real situations in the real world. Does computerization automatically make for development? Should it always? Who, from what countries, says it should or should not, and why? And what do they mean by development?

Doris Schoenhoff is deeply sensitive to these and related problems. She has worked as a computer specialist in countries throughout the world, many of them, such as locations in Africa and Latin America, possessing limited computer equipment, software, and trained users, and at the same time immersed in a great sea of often highly serviceable indigenous knowledge.

Computers have developed at the end of a long noetic tradition tracing back largely to the ancient Greek formalization of knowledge by a logic which encouraged the view that truth is maximized in propositional statements. (An alternative persuasion might be that truth is maximized in human living and/or in deep personal relations.) Ultimately, from the eighteenth century on in the West, the Greek-grounded view would help bring about the establishment of (logical) Western science as a practical guide to living, and, further, would tend to equate the development of human society and even of the human person with technological (logical, scientific) innovation. Computers enhance the promise of salvation through formalization of knowledge and technology because they not only store but, in a sense, breed knowledge outside the human mind, taking over, seemingly, a previously distinctively human activity.

There are even those who feel that computers can, ultimately, overtake or even surpass the human mind, substituting for it computerized Artificial Intelligence — although most researchers in Artificial Intelligence are quite aware that computerization can only approach human intelligence asymptotically, coming closer and closer to it (really, only to certain aspects of it) but in such a way that the closer and closer you get, the clearer it becomes that you are never going to get all the way.

In the indigenous, often basically oral, cultures of the Third World, as Dr. Schoenhoff shows, such formalized and abstract models of knowledge systems and human life management appear unreal, however the individual cultures may be impacted by such technological creations as television. Persons in these indigenous, basically oral, cultures are often strongly attracted to such technological creations of high-tech culture, without having either the acquired personality structures or the noetic or other defenses to cope with the cultural matrix which produces the creations that fascinate them.

But the indigenous cultures of the Third World are not of necessity totally subservient to high technology. They often have knowledge systems providing insights and patterns of living which are missing in cultures founded on high technology and which appear unmanageable as computerized knowledge. This is the seeming impasse to which Dr. Schoenhoff addresses herself, providing, if not a complete solution to the evident problems, some rich suggestions for working out solutions. Her basic suggestion is to attend to the ongoing debate as to whether Western science provides an entirely "objective" view of reality or is itself a particular view, not necessarily entirely false, but shaped by its own prejudices. Other suggestions include using insights developed in cognitive psychology, cognitive anthropology, and other disciplines managed in cross-disciplinary fashion; supplementing the economic and political focus which technological assessments and development strategies typically foster with an epistemological and ethical focus, especially with an ethical view of development; bringing in the world views of the poor and the illiterate not simply to correct such views — for the poor and illiterate must know some things about their existence which outsiders do not know — but to learn more profoundly by attending to the views of such persons how to interface computerized and noncomputerized, indigenous knowledge.

This work is in no sense an indictment of computerized expert systems, but it advances more than a hint of their possible limitations. Dr. Schoenhoff does not undertake to provide fully shaped models for imitation as computerized expert systems are brought to bear on indigenous knowledge systems, but she insists that it is fatuous to believe that computerized knowledge systems alone will solve what is essentially a problem of interfacing them with something much older and quite other than themselves.

This work is ultimately about conscious human communication, not about simple "information." Information is one thing, conscious human communication is another. Information systems need not be conscious at all. Billions of information systems, such as those encoded in DNA, have worked in each human being every moment of each individual's life without any human being's awareness — until recently — that they were there at all. Conscious human communication normally involves information, but it is not simply the movement of information from one person to another. Conscious human communication is an activity rooted in personal encounter between reflectively conscious beings, paradigmatically an "I" and a "you" (singular sense, as in the now archaic "thou"). As children, we all learned to communicate consciously as we learned, each of us, to set off his or her unique "I" through encounter with a "you." This kind of conscious personal encounter is not enframed in computer language. Computer language is enframed in personal encounter. People use computers. Computers do not use people. The interface between computerized knowledge systems and indigenous knowledge systems involves, in its depths, individual people and their life-worlds. Dr. Schoenhoff's basic and deepest message is that in the problems she deals with in both kinds of knowledge systems: it is people who are ultimately involved.

Besides its practical human applications, however, this book has even wider implications. Over the past decades massive scholarly work, including much fieldwork, has been done on the ways in which writing transforms the original oral mentality of humankind into the mentality of writing cultures, which is later transformed into the mentality of letterpress print cultures, which is still later transformed into the mentality of electronic cultures. Electronic cultures are marked by the "secondary orality" of radio and television — "secondary" because radio and television are the products not of humankind's original "primary orality" (the orality of cultures without any awareness of even the possibility of writing) but are, and can only be, products of cultures saturated with writing and print. But electronic cultures go further than generating a new orality. They give rise, through the computer, to a kind of secondary textuality, marked often though not always by planned interaction of verbal text with graphics, of the originally oral word with stark visual information. Writing and print textuality is no longer the ultimate contrast to primary orality. Electronic implemented verbalization is. Hermeneutics must reinterpret itself. In the history of verbalization, the digitized text as of now has some sort of latter-day primacy (which does not mean monopoly). This primacy of the digitized text many persons, including myself, have adverted to but never in the kind of detail that Dr. Schoenhoff manages here. That does not mean that digitization has the field to itself: digitization must react with the nondigitized, not single-mindedly and reductively, but openly and with more sophistication than ever.

Walter J. Ong
St. Louis University

Walter J. Ong, SJ, is a University Professor Emeritus of Humanities at St. Louis University. His more recent books include *Orality and Literacy*, as well as *Fighting for Life: Contest, Sexuality, and Consciousness* and *Interfaces of the Word.*

Acknowledgments

I would like to extend special thanks to Eugene B. Shultz, Jr., for his encouragement and suggestions during this project — even though I may have rushed where the proverbial angels would have buzzed the field. Through it all, he has been both mentor and friend.

Several other colleagues and friends read the manuscript for this book at various points of its completion, and I am grateful to each of them for their comments and criticisms, particularly Mark Rollins and Walter Ong. I also appreciate the efforts of Otieno Ndong'a to bring this book to the attention of readers in Kenya.

I want to recognize as well the clerical assistance that Amanda Velázquez frequently provided to me in the preparation of initial versions of this manuscript — often late at night and racing deadlines.

There are some debts that are too deep to articulate in a public space and are left unspoken in the heart. To my family, especially my mother, Anna Duff, my brother, Roland, and my sister, Mary Ann, to those whose friendship reaches back even to childhood, and to newer friends, such as Kevin O'Higgins, who have been supportive of my work by entering into a play of ideas with me, I can only return gratitude and love.

Finally, my thanks to Clement Burghoff for making my joy in seeing this book to publication his joy as well, and for always reminding me that it is "really wonderful" to be a writer. This particular writer, it needs to be said, benefited greatly from the good sense and good humor of her editor at Greenwood, Jude Grant.

Introduction

> Anything that you hear about computers or AI should be ignored, because we're in the Dark Ages. We're in the thousand years between no technology and all technology. You can read what your contemporaries think, but you should remember they are ignorant savages.
>
> — Marvin Minsky

This book comes with a warning label — as perhaps all books should. With my human ignorance noted and acknowledged, I will proceed through the darkness because that is the only course for those who grope toward the truth. There will be no pronouncements in these pages but only ideas — sometimes fragmentary, sometimes fearless. The nature of the subject demands both. It is complex and, as yet, lightly surveyed. If we waited for absolute certainty before exposing our thoughts on new issues, it is absolutely certain that no books would be published, and the human species would likely have slipped into the tar pits long ago.

JAGGED INTERFACE

The chapters to follow are about a particular interface. An interface is a common boundary. It is the site at which whatever exists on each side of that boundary — computer systems, equipment, human persons — can interact and communicate. Because it is a boundary with only a locus of interaction, the communication is never complete. Differences remain on either side. But changes also come about on each side. An interface requires mutual accommodation.

When the interface is between a Western technology, such as computerized expert systems, and a Third World environment, there will never be a perfect fit. If we propose to introduce such technology into the Third World, it is essential that we understand where the jagged edges of that interface will be and where the interface may fail entirely. That understanding may help us to predict some of the social and cultural changes that the new technology will propel — or, at least, to recognize them when they come hurtling down the tracks at us.

A CHIP OFF THE OLD EXPERT: MICROS AND MAESTROS

The technology of expert systems is one area of applied Artificial Intelligence (AI). DENDRAL, a program constructed to help chemists infer the molecular structures of complex organic compounds, is generally acknowledged as the first expert system. Consisting of about 400 rules, it was begun in the mid-1960s by Edward Feigenbaum, Joshua Lederberg, Bruce Buchanan, and other members of the Stanford Heuristic Programming Project (Buchanan and Shortliffe 1984, 8). The process of constructing an expert system is often referred to as knowledge engineering. It is a process of determining how decisions are made in an expert's specific "domain" or field of knowledge and attempting to symbolically represent that knowledge so that computer decisions can be generated in the manner of the human expert.

It is only recently that the use of expert systems in developing nations could even be considered. The reason is that prototype expert systems required large mainframes to execute. In the mid-1980s, there was a convergence of microcomputer technology and expert-systems development. It is this technological convergence that has made expert systems relevant for the developing world.

I emphasize the point that expert systems are computer programs. They form a new relationship between knowledge and technology. In *Electric Language*, Michael Heim detailed how electronic word processing brought about a different relationship to symbols (1987, 140). Microsoft Word, WordPerfect, and WordStar are not simply more efficient implementations of the technology of the typewriter. Electronic word processing allows for a manipulation, a networking, and an exploration of linguistic symbols that together are transforming thought and language in ways that print technology did not.

Heim suggests that with word processing "writing becomes more interactive and the computer becomes a place to think things out" with more spontaneity and without the constraints of linearity (1987, 153). Some critics insist this leads to fragmentation and repetition within the text. Of course, writers have always looked upon the blank page, whether in a typewriter carriage or on a note pad, as a "place" to think things out, to get their thoughts "down on paper." But the interaction of word processing is of a different order. The writer interacts not only with his or her terminal screen through a software package (instantly deleting, cutting, pasting, saving, or publishing the "text") but, very often, with geographically remote data bases, structured by other minds, and with a whole community of other users on the network. "There is not only a new technology available in word processing," Heim proposes, "but a gradually emerging sense of a new kind of community" (1987, 164). The activity of writing, and not simply the text, becomes more public.

In a similar way, expert systems are not just traditional methodologies that are stored now in software instead of in the "wetware" and textbooks of the human experts. While there is no general agreement about what constitutes an expert system, there are certain expected features of the software. These include the ability to draw inferences from a knowledge base and to justify the conclusions.

Expert systems are distinguished from the more familiar forms of information processing in three areas. First, they use a knowledge base rather than a data base. The distinction is perhaps a subtle one. The knowledge base consists of a flexible set of "facts." These facts and the relationships between them can be introduced and altered in an ad hoc manner. It is this lack of rigid structure that distinguishes a knowledge base from the more conventional data base in which data fields and relationships are predefined. Second, expert systems are generally written in special AI languages, for instance Prolog, that are suitable for a wider range of logical operations. And finally, expert systems attempt to solve problems through a symbolic representation of expertise and through various logical formalisms, such as predicate calculus, rather than through algorithms.

Just to nudge some memories, an algorithm is a precisely defined set of rules or procedures that is guaranteed to lead to a predefined goal in a finite number of steps. It often consists of a repetition of a sequence of operations, such as the simple procedure that schoolchildren used — before computers and pocket calculators — to find the square root of any real number. Computer programs that calculate payrolls, budgets, lunar landings, and the therapeutic doses for radiation treatment, all use algorithms, some of which may be part of special circuits that are printed on semiconductor chips. Such programs are quite different from expert systems. There are rules in expert systems as well, but they are rules for drawing conclusions from information that is often incomplete or approximate.

ARCHITECTURE OF THE MIND: SYNTAX AND SYNAPSE

Expert systems can be implemented, moreover, through a production architecture or a connectionist architecture, although this latter approach is currently very experimental (Gallant 1988). The production-system model of mind basically says that there are representations or symbols stored in human memory, that there is a set of if-then rules for manipulating these symbols, and that cognition consists of retrieving these stored representations and rules and performing logical operations. Production systems originated in the work of Allen Newell and Herbert Simon. Both men pioneered the computational model of mental processing.

In the production-system model, then, knowledge is represented by a set of syntactic and semantic conventions. The syntax of the representation specifies the rules for combining symbols into an expression. The semantics of the representation specifies how the constructed expressions should be interpreted. Knowledge representation is, in the production-system model, a language.

The connectionist model rejects the idea of stored symbols and logical rules and proposes a different architecture of the mind. Knowledge is not a matter of storage and retrieval; it is rather a matter of construction and reconstruction of mental representation within an existent situation. Knowledge representation in this model is a pattern.

Instead of one processing unit, there are hundreds or thousands of units in a connectionist system that are linked in a neural network, analogous to the neurons and synapses in the brain. There are different interpretations for the actual processing units in the connectionist model. In perhaps the most radical approach, mental representation is not in the output of a single unit but in the overall pattern of the network, which is the reason the system is also referred to as parallel distributed processing (Rumelhart & McClelland 1986).

In other words, knowledge in the connectionist architecture is not encoded in addressable locations; what the system "knows" is determined by the pattern of connectivity. It is the connections and the strengths of the connections that allow a pattern to be re-created. There are rules that govern the alteration and excitation of these connections and processing units so even connectionist systems must have rules.

PROPOSITIONS AND PATTERNS

While production-system models are largely concerned with the derivation, affirmation, and negation of propositions, such as if *p* then *q*, perception, not logical reasoning, appears to be the bedrock of cognition in the connectionist approach. Pattern recognition is the paradigm task studied by connectionists. It has proved difficult so far to develop rule-based systems with pattern-recognition capabilities because the machine must deal with a relevance that changes from context to context. In general, however, even connectionist systems have only been able to accept an input pattern or partial pattern, reproduce it or reconstruct it in a probabilistic manner, and then output that pattern.

Within AI research, pattern recognition now appears to have more applications than just visual or speech recognition. The whole process of categorization has been compared to pattern recognition in that it demands the ability to recognize similar items and group them together. Eleanor Rosch (1975) challenged the rule-based perspective of categorization by arguing that categories have a prototype structure in which some "exemplars" are judged to be better examples of the category than others; for example, a sparrow is a better exemplar of the category *bird* than a penguin. Such a prototype theory of categorization could more readily be implemented in a connectionist system in which the new input pattern would be matched against the previous patterns.

Likewise, Stephen Kosslyn (1977) and others have proposed that mental representations are not always linguistically based, but that there are functional images which are quasi-pictorial. Theoretically, these images are generated, not retrieved. If this is true, functional images also would be more readily adapted to a connectionist model.

THE WORLD INSIDE THE MIND

Both connectionist and production models, therefore, must wrestle with the question of what representation really is. Like the human mind, computers manipulate representations of the world. These representations are not a copy of sensory input because what is in the world is not identical to what reaches the senses. In the human retina, for example, there are an estimated 150 million photoreceptors, sensitive only to wave-lengths of light between 400–700nm, and one million optic nerve fibers.

Sensory input is filtered by the neurophysiology of the eye itself — by the number and kind of photoreceptors and by the chemical and electrical energy of the receptors and nerve fibers. Because of the proportionate difference, there is also considerable convergence of the photoreceptors on the optic nerves which conduct the electrical signals to the brain and excite cortical neurons.

At the cortex, these electrical events are further transformed and recoded. That recoding process within the human brain leaves out some of the details. There are many questions about what details are left out and about the form of the final code. Some think it is a linguistic code, and others a pictorial code. Still others reject the idea that mental representations are either language symbols or images, although they will accept a code that is nonlinguistic and holistic. Rule-based and connectionist models will not answer the questions about cognitive engagement in the same way. Much of the research into Artificial Intelligence, therefore, is focused on knowledge representation and knowledge acquisition.

THE GRAY REVOLUTION

At first, it would seem an incongruous match — Artificial Intelligence and Third World nations. Presently, there are few individuals even thinking or writing about the possibilities here. However, the discussion has at least begun. The National Research Council comprises members from the National Academy of Sciences, the National Academy of Engineering, and the Institute of Medicine. A 1986 study by an ad hoc panel of the National Research Council alluded to the potential for computerized expert systems in the developing world:

> Clearly, expert systems represent one avenue by which developing countries could free themselves from dependence on outside expertise, when such expertise is in short supply within the country itself. Although the level of reliance on knowledge-based systems is uncertain at this time, the systems could represent an important supplement to more conventional information resources. (1986, 186)

Two years later, this judgment was reiterated by another ad hoc panel of the National Research Council — this time with an enthusiasm rivaling even the market hype of some within the AI research community itself:

> A Gray Revolution is now beginning. Human reasoning — the use of the brain's gray matter — is being simulated by computers in a field broadly defined as Artificial

Intelligence (AI), which introduces new techniques to address development problems. Past progress in computer use and power, although rapid, has been evolutionary. Current and future applications of AI are revolutionary. (1988, 19)

Expert systems are relevant to developing nations because effective and ongoing education and management are critical issues in these environments. Many developing nations are clearly anxious to use this technology as a means of catching up with the industrialized nations. Analysts may continue to argue over how and where and when computer technology, including Artificial Intelligence, should be introduced into the developing areas of the world; however, its introduction has already begun and its further dissemination seems inevitable. At the University of Edinburgh, for example, there is now a group attempting to set up a resource data base on all known AI systems in the developing nations and to circulate information about upcoming AI conferences in places such as Nigeria and Cameroon.

TECHNOLOGY TRANSFER: THE THREAT AND THE PROMISE

While expert systems are currently being used for a variety of design tasks, scheduling, engineering and scientific applications, they are still at an early stage even at the large corporations, universities, and research laboratories in the Western nations. The most obvious means by which they will continue to find their way into Third World environments, therefore, will be through technology transfer. The ongoing expense, time, and technical experience required to design and implement expert systems will generally make the acquisition of off-the-shelf software and hardware from Western nations appear more practical than local development.

The introduction of expert systems, accordingly, is part of the larger question of technology transfer and the uses to which that technology is put. There is a fear that computer technology in the developing world will be just another tool for the elite — a fear that only a few will have knowledge of the technology, only a few will have access to the technology, and only a few will control the information the technology will make available. There is also a fear that the introduction of this technology will further distort cultural values in the developing nations, since technology effectively emerges from cultural settings and, in some sense, reflects the values of the cultures in which it was developed. And finally, there is a fear of further technological and economic dependence upon the West once the technology has been transferred.

Fear and apprehension, of course, have always shadowed the introduction of any new technology, even in the highly industrialized societies of the West. Just ask an American auto-worker about the assembly-line robots that turn out the Fords and the Chryslers and the Chevrolets these days, not to mention their even more prevalent and productive counterparts at Honda, Toyota, and Mitsubishi. Moreover, blue-collar workers and corporate CEOs alike feel a similar disquiet for the large computer data banks of personal information that the IRS, Social Security, VISA, AT&T, and dozens of other organizations and companies amass routinely on a daily basis. Any technology, in any environment, at any time has the potential to be misused or abused by those who have the power and the presumption to do so.

Partly for that reason, Western analysts stress the need for strategic planning to introduce computer technology into the Third World. However, one problem with these seemingly rational approaches to national planning and policy is that the development and the direction of a technology are not entirely rational processes. Intel Corporation, for example, did not set out to develop a microprocessor in 1971; they stumbled upon it. The computer vendors did not set out to create a personal-computer market; the technology created its own consumers.

AUTONOMOUS TECHNOLOGY

In the case of computer technology, we may be faced with what Langdon Winner calls "autonomous technology." The phrase and the concept are controversial. Winner is very aware of this himself. He concedes that some of his critics equate autonomous technology with technological determinism, which is "the idea that technological innovation is the basic cause of changes in society and that human beings have little choice other than to sit back and watch this ineluctable process unfold" (1986, 9–10). Winner goes on to say that such a notion would be too strong because there are real choices that arise in this process of social and technological transformation. With tongue-in-cheek perhaps, he suggests an alternative notion, that of "technological somnambulism," because, he remarks, "we so willingly sleepwalk through the process of reconstituting the conditions of human existence" (1986, 10).

The concept of autonomous technology may not apply to individual technologies but to the technological impetus in general. Or it may be more appropriate for computer technology rather than for mechanical technology. It does seem, however, to resonate a certain truth.

The subject of technology and modern technological society has generated a great deal of self-reflection and scholarship from writers, such as Martin Heidegger, Jürgen Habermas, Jacques Ellul, Lewis Mumford, Herbert Marcuse, Langdon Winner, Walter Ong, and many others. But the truth is that we still do not really understand technology in any depth — what impels its development or how it transforms our environments and our consciousness. This is one reason, perhaps, that technology often appears to be autonomous, even threatening.

Despite its critics and its limitations, the concept of autonomous technology does seem to have a certain validity, particularly if we are looking at technologies for electronic communication. Computers, like television, fall into this category. These technologies have already spread to some of the poorest areas of the world. It is not unusual to walk past the tin and cardboard shelters in the worst city slums of Africa or Latin America and spot a television aerial, beaming Western sitcoms, melodramas, and commercials. Personal computers have not reached this level of dissemination in the Third World. They have not quite become a mass technology even in the United States, despite early predictions that personal computers would be as prevalent in American homes as the television set. Nevertheless, they can readily be found in university, research, and corporate settings of Third World nations. As a result, the issue becomes not how to introduce the technology strategically into the Third World, but how to understand it, how to use it effectively, and how to prepare for the changes that will necessarily follow from its further dissemination.

LOCAL COMPUTER INDUSTRIES

This is not to deny that some Third World nations already have computer industries and universities with computer science departments and that individuals within these environments are working on designs for expert systems. Certainly, recent books and articles ensuing from international computer conferences document this interest of Third World nations in computer technology, including expert systems.

The fact that an expert system is designed within a Third World nation is, in itself, however, no guarantee that the system will reflect the knowledge, values, and aspirations of the local communities. Some of the local computer personnel may have been born or educated in Western nations. Some may have adopted Western attitudes toward technology or

Western values in general. As a result, they themselves may discount or disparage local knowledge and technology.

THE MATRIX OF KNOWLEDGE

It is my contention that computerized expert systems, which ostensibly encode knowledge and rules for decision, can never be completely extricated from the language, culture, and context in which they are designed and implemented. To be humanly meaningful in Third World environments, they must incorporate indigenous knowledge and indigenous reasoning about knowledge and contribute to internal solutions for development that are consistent with the cultural, social, and ethical aspirations of communities within the developing nations, especially marginalized communities. These knowledge-based systems can potentially be useful tools in developing the critical attention of community leaders because the system design process forces individuals to think about their knowledge and not just act upon it.

INDIGENOUS KNOWLEDGE

To insist that expert systems incorporate indigenous knowledge is to say that, in effect, there must be an ongoing interface between the computerized knowledge systems and the indigenous knowledge systems, between the system designers and the community. Computerized knowledge systems are those implementations of Artificial Intelligence that formalize and symbolically represent technological expertise. The term *indigenous knowledge system* has often been used in the anthropological literature (Brokensha, Warren, & Werner 1980). Within the context of this book, however, I am defining an indigenous knowledge system as the shared knowledge of a local community that has evolved in a particular Third World environment and that may be informally expressed in local custom, experience, technology, and wisdom.

In the much broader sense, of course, every culture has indigenous knowledge. For effective technology transfer, these two — the computerized knowledge system and the indigenous knowledge system — must be brought together. The challenge is to suggest where and how that interface may come about and to recognize the practical limits of both the indigenous knowledge and the computer technology.

The nature of this interface between the computerized knowledge systems and the indigenous knowledge systems is suggested in the title of this book. *The Barefoot Expert* refers, first of all, to the introduction of sophisticated computer systems into the "barefoot" environments of the Third World. While some of these environments may be very rich in terms of culture, social relationships, and human, ethical development, they are typically characterized by poverty in the economic sphere and powerlessness in the political sphere of international relations. Their poverty is defined in comparative national statistics by subsistence income, low levels of technology, high levels of illiteracy, political and economic instability, inadequate medical care, minimal opportunity, and a list of other economic and social indices.

The title also refers to the "barefoot experts" of the Third World — farmers, traditional healers, local environmentalists — who, I suggest, must participate in the design of these computer systems if the systems are to promote acceptable and enduring development within the indigenous communities. It is only arrogance that would argue that the expertise that matters for Third World development must come from the West, from universities, from multinational corporations, from international banks, from foreign and local professionals, but not from the farmer in Ghana or the healer in Botswana or the village teacher in Bolivia.

STRATEGIES FOR SURVIVAL

In fact, one of the unexpected outcomes of the environmentalist movement of the 1980s has been a recognition of the limits of technology to improve the human environment without adverse side effects, without diminishing the human spirit or threatening human existence. We are coming to realize that we must be concerned about the depletion of the rainforests, for example, not simply because of their natural splendor or because of a moral obligation toward all life — although these are sufficient reasons for many of us — but also because within the ecosystems of those rain forests are plants that might hold cures for human diseases, animals that might reveal secrets about sensory perception and adaptation, and, most importantly, other human cultures that have developed their own knowledge systems for survival. Though these may be utilitarian motives, they are generally the most persuasive in both First and Third World contexts.

Edward O. Wilson, a world-class biologist, acknowledges this even as he has attempted over the last decade to develop an environmental ethic. It is not surprising that most of us are fascinated by television documentaries on snow leopards or dinosaurs, that we seek out zoos and gardens in our cities, that we enjoy stroking migrating gray whales, listening for beast voices, or watching an old box tortoise make its way to the road's edge. Coining the term "biophilia" to describe a natural affinity of the human mind "to focus on life and lifelike processes" (1984, 1), Wilson proposes this life-bond as a new premise for a conservation ethic. He admits, however, that such an ethic must finally be grounded in selfish reasoning: we will conserve the land and the diversity of its species only if there is material gain for ourselves, our family, and our "tribe."

In the past, we have so easily dismissed all these environmental concerns as unimportant in the age of million-dollar laboratories, synthetic chemicals, test-tube life, and other shining technologies. We have looked upon many Third World cultures with a mixture of curiosity and cockiness. Because their rituals and practices were "unscientific," they were often judged to warrant little else. The Dogon of Mali, for instance, still invoke spirits and offer animal sacrifices. Dogon diviners trace symbols in the red sand at sunset and wait for the night fox, which they believe to be an oracle, to leave its pawprints in the sand — prints that the diviners will interpret at sunrise. The Surma of Ethiopia paint their naked bodies with chalk designs and their women wear large clay plates in their outstretched lips. But, amazingly, the Dogon have survived for generations on rugged slopes of scree, lashed by the harmattan winds, and the Surma have endured on their desolate lowlands with a simple diet of milk and blood from their treasured cattle.

Granted, the goal of humanity is more than simple survival. However, such cultures as the Dogon and the Surma have a rich knowledge of their environment. This is almost a requisite for continued existence in the generally low-tech cultures of the Third World. David Lewis recognized this when he spent nine months in the late 1960s sailing with Polynesian navigators who used the old star paths of their ancestors. He writes in *The Voyaging Stars* about these mariners who, with skills that have existed for over 4,000 years, sailed, in open canoes and without instruments, for thousands of miles by reading the stars, interpreting wave patterns, cloud formations, ocean luminescence, and the presence of sea birds. He also studied the remarkable land-navigation techniques of the aborigines as they guided him, without a compass, through the Great Western Desert of Australia.

Moreover, even the most remote Third World societies have an awareness today of Western culture and the values that are seen in Western cities. The Kogi, with a civilization that predated the arrival of Columbus in San Salvador, invited Alan Ereira, a film maker, to visit with the leaders of their community of 12,000 people in the Sierra

Nevada de Santa Marta of Colombia. They had withdrawn from outside contact 400 years ago, and today even Colombians cannot speak their language. Ereira, accepting the Kogi's one-time offer to hear their message, produced a fascinating documentary in 1991 for BBC Productions called "From the Heart of the World: The Elder Brothers' Warning." The Kogi look upon themselves as the "elder brothers" of the world who must teach their greedy "younger brothers" to take care of the earth so that it will not be destroyed. The film made it strikingly clear that, in one sense, the Kogi have much to teach our Western societies that often pay little more than lip service to ecology. For the Kogi, ecology is both survival and morality — as it was for the Indian nations of North America before their lands were changed forever by the European settlers. And, in spite of their rejection of Western behavior, the Kogi still have a sense of human solidarity which leads them to call us their younger brothers.

RAINFOREST LABORATORIES AND FOLK HEALERS

It is just as important, of course, not to romanticize Third World cultures. Living in a Third World environment does not make you a born ecologist. The greatest diversity of plants and wildlife on the earth today exists in the tropics. There are an estimated 250,000 species of vascular plants on earth, and two-thirds of these occur in the tropics and subtropics (Davis et al. 1986, xxxv–xxxvi). This is also where the majority of underdeveloped nations are. The pressures of poverty, famine, political instability, and war, often coupled with a lack of understanding about ecological issues among the populations of these nations, have created conditions in which animal and plant species are being destroyed at a chilling rate. Botanists, wildlife biologists, and conservationists from around the world have sought not only to save these animal and plant species, but to catalogue and analyze the existing knowledge about these species using the techniques and resources of Western science. No one would want to diminish the knowledge and the dedication that this effort has required.

Because Third World nations are primarily agrarian, much of the knowledge of Third World cultures concerns plants and agriculture. Of all the earth's plants, only about twenty of them provide 85 percent of the world's food source and only a few hundred are widely cultivated (Davis et al. 1986, xxxvi). By 1980, it was estimated that approximately forty-one species of plants accounted for almost all the plant-derived drugs in the United States, such as digitalis and digoxin, which are both from varieties of the foxglove (Myers 1983, 91). With so many species untested and even unidentified and unknown, there is a new research thrust to learn about the medicinal and nutritional uses of plants in many of the remote areas of the Third World. Not only the plants themselves but the soils as well are crucial. The microorganisms in these soils are a main source of our antibiotics and other medicines.

All you need to do is pick up an issue of *Economic Botany*, for instance, to find new research and articles on folk medicine. The Guarani Indians of Paraguay, it is reported, may use as many as 1,500 species of plants for healing remedies (Basualdo, Zardini, & Ortiz 1991, 86). In China, about 1,700 plants are in common use as medicines, and in India the number is about 2,500 (Myers 1983, 93).

The traditional healers in the Third World do not know the chemical constituents of the plants, but they know which plants to seek out to reduce bleeding during childbirth, to counteract amoebic dysentery, or to lower fevers. The Chinese were using a mold from tofu to cure boils and other infections 2,500 years before Alexander Fleming discovered penicillin. Even today, perhaps as many as 80 to 95 percent of the population in Africa depend upon herbal medicine (Joshi & Edington 1990, 71; Myers 1983, 94).

In many cases, there is not a system for passing on this knowledge and the knowledge itself is often disparaged by the younger generations. Alex Haley, who turned his grandmother's front-porch stories into the international best-seller, *Roots*, said that when an old person dies, it is like a library burning down. This is urgently true in the residually oral cultures of the Third World. The study of folk medicine and medicinal plants, therefore, is not simply a curiosity for anthropologists and ethnobotanists. It is becoming increasingly more important to Western pharmacologists and chemists, as well. Ephedrine, for example, is a decongestant that is used in many of our nasal sprays and cough syrups. It comes from a semidesert shrub, Ephedra, that has been known in China for 5,000 years (Myers 1983, 92). Western scientists can only wonder how many other "undiscovered" drugs are right now being used effectively by the traditional healers of the Third World.

A PENNY FOR YOUR THOUGHTS: THE KNOWLEDGE MARKET

The danger, however, is that we will seek this knowledge primarily to improve the quality of life in the West, that we will extract this knowledge as we have extracted the natural resources of the Third World to fuel our own development. According to a 1990 study from Tufts University, it takes, on the average, almost twelve years and $231 million to bring a new drug from the laboratory to the marketplace in the United States (Gordon & Wierenga 1991). Many of our newer drugs are biogenics that act upon existing biological processes in the body. Only about 10 percent are preparations from healing plants (Dörfler & Roselt 1989, 11). So we are obviously not talking about snatching wonder plants out of the hands of poor Third World healers and returning to the West to make an instant fortune.

Although pharmacological testing of healing plants has in many cases confirmed knowledge that has existed for thousands of years, it is also difficult to determine the active substance of a plant that appears to produce a medicinal effect (Dörfler & Roselt 1989, 11). Admittedly, though, both academic reputations and commercial products can be fostered from this traditional knowledge.

It is extremely difficult to define the ethics that should come to bear upon Western researchers working in the Third World. Some would suggest offering a remuneration to those who gather the plants and pass on their healing knowledge. But how much should the remuneration be? If it is too high, it will drastically change the existing society and culture. If it is too low, it will put the researchers in the position of exploiters. Others would deny that any remuneration is necessary or desirable, unless requested, since it forces the same values on the traditional society that exist in our industrialized societies. We assume that everything has a price. With either approach, the threat of exploitation is real.

MYTHINFORMATION

This threat of exploitation is intensified when we begin talking not simply about the knowledge that single individuals within a culture might have about local plants or techniques, but about the whole cultural context of knowledge itself and the impact that Western computer technology might have upon a Third World environment.

Langdon Winner, for one, has labeled many of the deep-rooted beliefs and exhilaration surrounding all the discussion about the computer revolution as "mythinformation." Like myth, Winner argues, the talk about great information societies contains an element of truth, but you have to look at who is benefitting from all these changes. "Current developments in the information age suggest," Winner says,

> an increase in power by those who already had a great deal of power, an enhanced centralization of control by those already prepared for control, an augmentation of wealth by the already wealthy. Far from demonstrating a revolution in patterns of social and political influence, empirical studies of computers and social change usually show powerful groups adapting computerized methods to retain control. That is not surprising. Those best situated to take advantage of the power of a new technology are often those previously well situated by dint of wealth, social standing, and institutional position. Thus, if there is to be a computer revolution, the best guess is that it will have a distinctly conservative character. (1986, 107)

Certainly, what limited discussion there has been so far about expert systems for Third World development has focused on the benefits these systems will bring to academics, professionals, administrators, and managers within the poor nations. And, in one sense, all that is necessary because, if these well-trained individuals were not part of the existing knowledge pool, no one would even consider introducing expert systems into a Third World environment.

Although competition is a very American value and, arguably, a very destructive one, it would be unrealistic to deny that there is pressure — from without and from within, from society and from individuals — for scientists, technicians and students in Third World nations to stay current with scientific and technological developments in the industrialized countries. One could argue endlessly about whether this desire to emulate at least selective aspects of the Western lifestyle comes from the colonial aftershock, whether the West has created needs in the Third World that did not exist before, or whether Third World nations should take a noncompetitive approach to development. But the competitive urgency is clearly there.

When we are considering potential applications for expert systems, we, therefore, tend to look at areas such as medical diagnosis — first, because of the perceived need to improve medical care in Third World countries and, second, because we assume that modern medical science should be universal in the United States, Kenya, or Paraguay. Science is science. There is a degree of truth to that — which also implies that there is a degree of untruth. Medical science is not simply the latest technology for laser surgery or for microchip implants. It is also about practice and attitudes, and these will be as diverse as the cultures themselves.

PULLED UP BY THE ROOTS

What I propose is not to ignore these sophisticated types of applications but simply to expand the focus of any discussions to include the poor. We cannot simply concern ourselves with the benefits that expert systems can bring to a few in these poor nations — the few, for example, that work in a modern medical facility or that can afford to be treated there. We also have to look at what benefits these systems can bring to the poor majority in these nations and what additional threats they might pose to an already marginalized existence.

Admittedly, suggestions for popular or grassroots participation often meet with the offhanded criticism that the proposers of such approaches are naive and unaware of the facts of social, economic, and political power. As compelling as that power might be, this insight remains clear as well: without grassroots participation in the design and use of this knowledge-based technology, there can be no lasting benefit and no empowerment for the poor within the Third World nations.

It is the needs of the poor that must set the primary agenda for development within each nation, precisely because their lives are what define lack of development in the cultural society and in the human society. That is a point that can be explained and defended, but not proved or disproved. It is a conviction that some bring to the debate about development and others do not. Those who do hold that conviction, I think, must state it straightforwardly.

ANALOGUES OF WISDOM

For my own part, I also believe that our lives require wisdom and not simply empirical research from laboratories. They require self-knowledge, self-possession, and self-sacrifice. These make possible good judgments and peaceful lives even though situations are complex, experience is limited, knowledge is uncertain, and death is inevitable. That is, in part, what the barefoot experts of the East — the Buddha, the Christ, the Prophet, the Mahatma — have taught the Western world.

I deliberately point to individuals and not to religious institutions as exemplars or analogues of wisdom. Each of us, no matter what our religious affiliation or lack of one, would have to acknowledge that religious institutions sometimes deviate from their models. There have been many personal and cultural atrocities perpetrated in the name of religion. In some cases, priestly castes and "mysteries" have been used to keep people under control or to exclude those who are not the right sex or the right color or the right ancestry from power and participation.

At times, religious pressure has been exerted to actually retard scientific knowledge and to promote faith over reason. It is at least an embarrassment to Catholic intellectuals that Galileo was compelled by the

Inquisition to stand trial in 1633 for his "heresy" that the earth moved around the sun. Addressing the Pontifical Academy of Sciences on October 31, 1992, Pope John Paul II formally proclaimed that the Catholic church had erred in condemning Galileo. Vatican "experts" had studied the case for thirteen years before presenting their conclusions to the pope (*St. Louis* 1 November 1992: 5A). By any standards, 360 years is a long time to wait for a retraction. Some see a comparable travesty of religion in the fact that American clergy blessed the bombs in World War II and, more recently, that George Bush assured his television audience that God and right were on America's side in the Persian Gulf War. Religion has often been used to justify political positions, from the holocaust of Nazi Germany to apartheid in South Africa, oppression in Northern Ireland, and the "ethnic cleansing" of Muslim Slavs and predominantly Catholic Croatians by the Serbs of Bosnia-Herzegovina.

Religious institutions, precisely because they are human institutions, have unfolding histories of their own and a gradual awakening to truth. More important than institutions are what the lives of Jesus, Gandhi, Buddha, Mohammed and others have to teach us — what values they actually lived. And certain values seem to be common to the lives of these individuals. Their lives are ethical models. While some might argue that we learn from experience and not from example, I think we learn from the experience of others and not just our own.

Above all else, development requires an ethical approach, some understanding of the nature of the human person and the responsibilities of one person to another, of the individual to society, of society to society, and of society to the natural world. This ethical basis has to come from the lives of these ethical models, from the analytical principles of philosophy — or from both. There are difficulties no matter which way you turn. While Gandhi was venerated by many Indians as a holy man, his way of nonviolence, self-discipline, and simplicity was not the road to economic development and political power that India chose. Some resented that Gandhi offered poverty as almost a means of sanctification or that he urged love and tolerance between Moslem and Hindu and the liberation of the untouchable caste. The model was rejected as too austere for anyone but a Gandhi to follow — just as the example of Jesus has often been ignored by Christians.

On the other hand, the idea of deriving ethical principles strictly from philosophical analysis might appeal to an educated elite but not to the ordinary individual. The people may rally to "Liberty, Equality, Fraternity," but their expectations from the revolution are much more immediate. This was clear from events in the republics of the former

Soviet Union, immediately following its collapse. People waiting in line for hours or even days to purchase bread or shoes, when these were even available, at prices inflated by 200, 300, or more percent, mumbled that they were better off under Stalin. Food became more important than freedom. It is no good if the ethical principles find their way into constitutions or human rights statements or textbooks, but not into practice. In practical terms, most of us are drawn to emulate human example, not textbooks. How to derive and promote a global, human ethic is, therefore, one of the thorniest issues of the development debate.

TRACKING A WOOZLE

Finally, when others become aware that you are doing research or writing a book, you eventually get asked, "What's it about?" Often, you can skip around such questions. But with persistent colleagues, you are forced to come clean. Invariably, the reaction to my own response that I was writing about expert systems in the Third World was one of amusement. When I was told enough times, with good humor, that expert systems obviously did not have any relevance for the Third World — and particularly the poor of the Third World — I knew I was on the right track.

Granted, some who read this book will feel that other authors could have been included. I would not dispute that. There is such a swell of scholarship and publication today that it is impossible to be conversant with all of it. Some might wonder why one author was selected over another. A few of the authors that I did select have informed my thought for many years. Others were chosen not because they were "in" or "hot" but because aspects of their work were compatible with my own intuitions and experience. I feel free to gather insights from those who write books, without necessarily subscribing to the whole philosophy of the author, as well as from those who cannot even read books, but are nonetheless wise. There are other teachers, too, hidden teachers — an eagle sailing above the frozen river in the winter light or a tiny spider, thinking spider thoughts and spinning a morning universe between blades of prairie grass. From those kingdoms not our own, I have also gathered treasures.

"Every man tracks himself through life," Henry Thoreau confided to us. By this he meant that in our observations, our reading, and our travels, we will only see what our eyes and our temperament permit us to see and hear what our ears and our heart allow us to hear. It is our own footprints we follow — sometimes in circles like Winnie-the-Pooh tracking a Woozle. I only hope my circles have been wide and far-ranging. Like Thoreau, another scout of the truth, I have simply reported back on my sojourns.

Chapter 1

Science under the Microscope

> I do not know what I may appear to the world, but to myself I seem to have been only like a boy playing on the sea shore, and diverting myself in now and then finding a smoother pebble or a prettier shell than ordinary, whilst the great ocean of truth lay all undiscovered before me.
>
> — Isaac Newton

> We have learned that nothing is simple and rational except what we ourselves have invented; that God thinks in terms neither of Euclid nor of Riemann; that science has 'explained' nothing; that the more we know the more fantastic the world becomes and the profounder the surrounding darkness.
>
> — Aldous Huxley

> The scientific theory I like best is that the rings of Saturn are composed entirely of lost airline luggage.
>
> — Mark Russell

Wonderfully at ease in both science and letters, Lewis Thomas once remarked that "the great thing about language is that it prevents us from sticking to the matter at hand" (1974, 95). Our matter at hand is the relevance of expert systems to Third World development. Right away, we must start tacking because all these terms — expert systems, Third World, development — refuse to hold a steady course. A certain amount of indirection and ambiguity is unavoidable when we begin discussing any subject of real importance. "We make detours," Ludwig Wittgenstein wrote about his own approach to philosophical investigation, "we go by side roads. We see the straight highway before us, but of course we cannot use it, because it is permanently closed" (1953, 127).

In fact, indirection appears to be quite essential to human understanding. After all, we say that we learn by "trial and error." This method of validating or refuting a hypothesis through repeated experimentation is the very core of what, since the seventeenth century, we have respectfully referred to as the scientific method. Interestingly, the word *error* comes from the Latin root *errare*, which originally meant *to wander about*.

I am going to take the liberty of wandering about in this chapter because the subject of expert systems and Third World development involves much more than just the introduction of a new piece of Western technology into a non-Western culture and the effect this technology will have on persons in that culture. It also involves us and our own attitudes, including a popular belief in the irrefutability and universality of science and applied science.

OFF TO A LATE START

From the fifth century BCE until approximately the sixteenth century AD, science was a part of natural philosophy. The early Greek philosophers, such as Leucippus, Democritus, Epicurus, Aristotle, and Archimedes, were the ones concerned with the nature of matter, motion, and multiplicity. At the time the University of Paris came into being around AD 1150–1170, there were only four faculties: arts, medicine, canon law, and theology. Physics was still simply a part of philosophy. In fact, through the Renaissance, physics, along with logic, comprised virtually the whole of philosophy at the universities; metaphysics and ethics were practically lost in the leaves of the scholastic commentaries. Like philosophy, physics was approached through the study of ancient Latin texts and rhetorical methods.

At least partly through the establishment of discussion networks, such as the Royal Society of London (1662) and the Académie des Sciences of Paris (1666), and also with the publication, in Latin, of Newton's *Principia Mathematica* (1687), science gradually became distinct from philosophy, although it was still closely linked with mathematics. But you could not practically speak about communities of physicists, chemists, and biologists until around the nineteenth century with the likes of John Dalton, Gregor Mendel, Friedrich Kekulé, Charles Darwin, Dmitri Mendeleev, Wilhelm Roentgen, Louis Pasteur, Antoine Henri Becquerel, André-Marie Ampère, Michael Faraday, James Joule, James Maxwell, and others.

Even though the nineteenth century produced these scientists and greatly expanded scientific knowledge, research was generally done outside of the university. Scientific knowledge was largely disseminated through elite white-male societies, through published texts, and through itinerant public lecturers (Jacob 1988). London University, for example, did not establish a faculty of science until 1859 (Morrish 1970, 93). So science, as we have come to know it, is a late arrival, though not as late as the study of vernacular languages and literature.

It is not an exaggeration to say that human society has been transformed by science and by the new technologies that have followed from scientific discoveries. Technology, of course, has not always derived from science. In some cases, science was the debtor to the backyard tinkers and inventors. "Science," Clifford Geertz aptly wrote, "owes more to the steam engine than the steam engine owes to science; without the dyer's art there would be no chemistry; metallurgy is mining theorized" (1983, 22). Interestingly, the term *technologia* originally meant the logical presentation of grammar and later of any subject and has migrated slowly to its definition today of applied science (Ong, [1958] 1983, 197).

OBJECTIONS TO OBJECTIVITY

We are well aware by now that there is a bias even to science itself. The scientist approaches reality with a preconceived method, and often, with a particular question in mind — at times even with an expected answer. When Gunnar Myrdal said that questions must be asked before answers can be given, he was suggesting that both our questions and our answers reflect our culture, our gender, our previous knowledge, our expectations, and our fears. "Despite the quasi-theological injunction, 'Thou shalt let facts speak for themselves,'" Pierre Pradervand quips, "no facts can do that" (1989, xvi).

For many, Michael Polanyi put the idea of scientific objectivity to rest in the late 1950s. If we were to look at the universe with true objectivity, Polanyi claimed in the opening chapter of *Personal Knowledge*, it would be like replaying a feature film of the complete history of the universe in which the achievements of humanity would flash by in the final few seconds. Similarly, if we paid equal attention to equal portions of mass, we would have to spend our entire lifetime on interstellar dust and it would take a billion lifetimes to get around to humankind — and even then we would warrant only a second's notice. "It goes without saying," Polanyi concludes,

> that no one — scientists included — looks at the universe this way, whatever lip-service is given to objectivity. Nor should this surprise us. For, as human beings, we must inevitably see the universe from a centre lying within ourselves and speak about it in terms of a human language shaped by the exigencies of human intercourse. Any attempt rigorously to eliminate our human perspective from our picture of the world must lead to absurdity. (1958, 3)

Scientific objectivity has come under attack from quite another front as well. Today, the practice and institution of science is under close scrutiny — like some threatening microbe fixed to a slide and waiting for a closer stare. In the last decade, it has been peered at and probed by congressional committees, university faculty, and investigative journalists. The cause of all this attention has been mounting reports of fraud and scientific misconduct in some of this country's most prestigious universities and research laboratories (Bell 1992; Broad & Wade 1982; Kohn 1986).

As a result, scientists and philosophers have put science itself under the microscope. Peter Medawar, a Nobel Laureate in biology and an articulate spokesman for the scientific enterprise until his death in 1987, wrote an essay entitled "Is the Scientific Paper a Fraud?" Medawar argued convincingly that the scientific paper not only conceals the reasoning process that scientists go through, but actually misrepresents what it is that scientists do (1990, 228). It tidies everything up — smudged notes in the margins, unlikely results, test-run chronology — and puts it all in a neat, inevitable order. Max Delbrück, one of the pioneers in molecular biology, even coined a phrase for the contribution that error sometimes makes to scientific discovery: "the principle of limited sloppiness." For his part, Medawar proposed that instead of marveling at the scientific frauds that are uncovered, we should marvel at the fact that attaching importance to truth is a rather recent innovation in human history, "if by truth we mean correspondence with empirical reality" (1990, 64).

PARADIGMS FOUND

Thomas Kuhn's headline of thirty years ago has also become a familiar item now in the scientific literature and thinking. Kuhn proposed that science does not develop by the progressive accumulation of individual discoveries and inventions but rather is directed by paradigms. Science and technology are shaped as much by these paradigms as by any scientific method. Information-processing psychology, for example, was the dominant paradigm in the 1960s and 1970s for studying cognitive processes (Lachman, Lachman, & Butterfield 1979, 6).

In effect, Kuhn says that it is a mistake to speak about progress in science as though the theories of the previous generation of scientists were simply adjusted and refined by the present generation. Scientific knowledge, according to Kuhn, is not incremental. This almost seems counterintuitive to most of us because we are so accustomed to equating new technology with this scientific progress. No one would question, for example, that the electron microscope has enhanced our ability to see. Today the microbiologist can actually stare at a virus and watch as it takes over a cell's production center, producing virus particles instead of cell protein. Through photography and television, this enhanced vision is brought right into the ordinary home. When we watch a program on "Nova" about human conception, it is virtually impossible not to be dazzled by the sight of a sperm fertilizing an ovum or of the early division of cells.

Kuhn would admit that the technology is better. He questions, however, what we see with that technology and to what extent our scientific theories limit and determine our vision. For Kuhn, a paradigm is a manner of seeing. When one way of seeing is superseded by another, the result is scientific revolution.

The usual prelude to a revolution in science and a change of scientific paradigms is crisis (Kuhn [1962] 1970, 181). This crisis, according to Kuhn, is not always generated within the scientific community. It can come from without the community itself and be the result, for example, of new instruments, such as the electron microscope or the electronic computer. When this crisis occurs, some individuals will no longer see the world as they had before. Relationships and groupings will have changed. As this happens, a new paradigm emerges.

Kuhn, in fact, used the term *paradigm* in various senses and admitted that he created "gratuitous difficulties" for his theory, which he began disentangling in response to his earliest critics (1970, 174). His fundamental viewpoint, though, has generally remained unchanged. On the one hand, a paradigm comprises the "beliefs, values, techniques, and so on shared by the members of a given community" (1970, 175). A scientist usually acquires his paradigm in graduate school, and his scientific beliefs are maintained by his membership in a scientific community of believers. The opinions of his colleagues dictate his career and determine whether he will be accepted and promoted. The opinions of journal editors dictate whether his research will be published. The opinions of evaluators on foundation committees determine whether he will receive funds for further research. There are subtle pressures to "fit in."

However, apart from this sociological sense of the term, as Kuhn calls it, a paradigm is also the knowledge which is "embedded in shared examples" (1970, 175). Kuhn even suggests substituting the word "exemplar" for "paradigm." Exemplars are "the concrete problem-solutions that students encounter from the start of their scientific education, whether in laboratories, on examinations, or at the ends of chapters in science texts" (1970, 187). Physicists study physics by learning the same exemplars. And so also with chemists, biologists and other scientists. Kuhn insists that the concept of the paradigm as shared example is the central element of his work (1970, 187).

The exemplary problems train the student of science to see situations in a certain way. Moreover, this training is partly learning the words of the scientific community that describe aspects of nature (mass, acceleration, friction, gravity) and partly being shown examples of how these words relate to nature. Defending himself against the charge that he has made science subjective, irrational, and based on intuitions rather than logic and law, Kuhn argues that the intuitions are communal and tested, not individual and opinionative.

In addition, the knowledge embedded in the shared exemplars is also systematic and analyzable but, Kuhn explains, "is misconstrued if reconstructed in terms of rules that are first abstracted from exemplars and thereafter function in their stead" (1970, 192). For Kuhn, one becomes a scientist primarily "by doing science rather than acquiring rules for doing it" (1970, 191).

CONSENSUAL SCIENCE: YOU SEE WHAT WE MEAN

The Structure of Scientific Revolutions was not only about the role of paradigms in scientific research but about the constraints of community as well. Kuhn described the socialization process of professional education and the molding of graduate students to the commitments, the terminology, and the methods of a scientific discipline. A few years after Kuhn's book was published, John Ziman, an English physicist, wrote about the nature of that scientific community. Basically, Ziman said that the goal of science was not to produce "published knowledge" but "public knowledge," which is a "consensus of rational opinion" in a particular field (1968, 9). This consensus is not logical but experiential, social, and contextual. The young scientist, according to Ziman, does not study formal logic; rather he "learns by imitation and experience a number of conventions that embody strong social relationships . . . he learns to play his *role* in a system by which knowledge is acquired, sifted and eventually made public property" (1968, 10).

One of the problems with consensual science is that the consensus can be wrong. An example of this is the rejection of Wegener's hypothesis of Continental Drift for over fifty years (Ziman, 1968, 56–57). Another problem is that if we define a scientific community in terms of the education and expertise of its members and define that education and expertise by the community, we are left chasing our "tales" in circles and the public knowledge seems to become less universal — less public.

In the intervening years, philosophers have had time to think about the theory of consensual science and some are having new doubts. Lately, Jerry Fodor has raised questions about "perceptual implasticies, cases where knowing doesn't help seeing" (1984, 34). He uses the example that even though we may know an optical illusion is an illusion, this does not change our perception. Fodor's example is not quite as straightforward as it seems because there have been some very interesting articles on perceptual illusions that indicate individuals from non-Western cultures may not perceive pictorial representations as we do (Deregowski 1989). Admittedly, these studies are themselves somewhat problematic since, as one researcher admits, they "have progressed by fits and starts and have been propelled by the efforts of individual scholars rather than by those of schools" (Deregowski 1989, 112). As a result, many cross-cultural hypotheses and conclusions are based on anecdotal rather than experimental evidence.

Fodor, nevertheless, is trying to make a point about the intractability of human perception. "If we ought to be impressed by the degree to which perception is interpretive, contextually sensitive, labile, responsive to background knowledge and all that," Fodor says, "we surely ought also to be impressed by the degree to which it is often bullheaded and recalcitrant" (1984, 33). Fodor is asking whether observation can be "theory-neutral," whether scientists with different theories can observe the same phenomena, whether there is one perceptual world or not. If the answer is yes, then, Fodor argues, it is observation and not consensus that compels rational belief.

If the scientific paradigm trains us to see and experience the world in a certain way, it is also true that those who do not share this paradigm in a sense live in a different world. Referring to the neural processing that takes place between what we conventionally describe as a stimulus and a sensation, Kuhn warrants "that very different stimuli can produce the same sensations; that the same stimulus can produce very different sensations; and, finally, that the route from stimulus to sensation is in part conditioned by education" (1970, 192). Education, language, experience, and culture, all create, according to Kuhn, the "objects" that populate our world. This hypothesis has clear implications for both scientific knowledge and the indigenous knowledge of other cultures.

While Kuhn has his own critics and would be the last to say that his position is fully worked out, his insights have changed the way the present generation of scientists look at science. We recognize, as even the earlier giants of science did not, that our prejudices are concealed in our definitions and conceptualizations. Science itself does not escape the enculturation process. There are several recent books that take up, at least in part, the history of Western scientific culture and the possibility or impossibility of scientific neutrality, particularly today in the face of the militarization and commercialization of science, the government funding of research, the industrial investment in science, and the conflict of interests of entrepreneurial scientists, notably in biotechnology (Jacob 1988; Proctor 1991).

REDUCTIONISM: FROM ONE MANY

Theories of observation and consensus aside, scientists have filled textbooks and libraries with information gained by a process known as reductionism. All along, Western science has sought to reduce reality to its most fundamental laws, its most fundamental particles. The object of the game was to work back from the complex to the simple and then derive from those fundamental laws and fundamental particles more complex laws and systems. In the process we had definitions for everything — quarks and black holes, mitochondria and neurons.

Meantime, our life-world, the world of our own past and present experiences and the experiences of others that we daily take into ourselves through language, was often not recognizable as the one we found neatly defined in textbooks about quantum physics, cell biology, neuropsychology, and a myriad of other scientific subjects. Most of us go about our entire lives never knowing much about those definitions and details — or scarcely wondering about the inner structures of cells or galaxies.

As it turns out, we have been much better at taking apart than we have been at putting together. We have generally overlooked the world with our preoccupation with its parts. We do have partial theories of the physical universe — the general theory of relativity and quantum mechanics. The general theory of relativity explains the force of gravity and the relation of space and time within the structure of the universe itself. Quantum mechanics describes the unobserved universe of subatomic particles.

In his book, *A Brief History of Time*, Stephen Hawking, who is generally regarded as one of the most brilliant theoretical physicists since Einstein, points out that "these two theories are known to be inconsistent with each other — they cannot both be correct" (1988, 12). The discrepancy between these partial theories was mathematically confirmed back in 1972 (Hawking 1988, 157). As yet, however, no new hypothesis has been offered to replace them or reconcile them. The Grand Unified Theory escaped Einstein as it has escaped Stephen Hawking today.

HOLISM: FROM MANY ONE

Still, the reductionist approach has, without question, advanced our understanding of the natural world and made possible all the shining, new technologies, such as lasers, space satellites, gene-splicing, and supercomputers. But even within the physical sciences, there is a growing recognition that a reductionist approach cannot completely explain the complexity of the world. Some scientists have grown book-weary of simply naming elements and particles. They argue that you cannot completely understand even the natural world by breaking it into smaller and smaller parts.

These include physicists such as David Bohm, whom Einstein once suggested might be the one to solve the unified-field theory (Briggs & Peat 1984, 144). Bohm attacks the assumption that nature can be analyzed into parts, and wants to substitute a holistic approach to the universe as opposed to a fragmentary one. He is not dismissing the specialized research fields or the conceptualization of physical reality into abstract parts. He is challenging the fragmentation that ignores the wider context and that assumes our quarks and mesons and leptons have a reality independent of our scientific theories and the instruments forged by our technology (Bohm & Peat 1987, 15).

Since, according to Bohm, nature is an organic whole and is not composed of parts, the patterns and relationships that we postulate when we separate the whole into parts do not reflect the real order of nature, an order which Bohm calls a hidden or "implicate order" (Peat 1988, 26; Briggs & Peat 1984, 114). The particle is not an independent thing at all. It is an abstraction for what is really enfolded into one reality.

The following example helps to clarify what Bohm meant by "implicate order" and "explicate order," by enfolding and unfolding (Zukav 1979,

309–10). Imagine that a small hollow cylinder is placed inside a larger cylinder. In the space between the two cylinders, a viscous liquid, such as glycerine, is poured. Afterwards a drop of ink is placed on the surface of the glycerine. When one of the cylinders is rotated, the drop of ink, a particle from the explicate order, spreads out into a thin line and finally disappears altogether. The drop of ink is enfolded in the glycerine and becomes one with the implicate order. When the cylinder is rotated in the opposite direction, the ink drop reappears and is unfolded from the glycerine and appears again as a part of the explicate order.

Although some have dismissed Bohm as proposing a mystical or New Age physics, others are at least sympathetic to his concerns if not his overall theory. There is a recognition among many scientists that the behavior of the whole is different from the behavior of the parts. The flocking behavior of birds, for example, can never be explained by looking at the anatomy and behavior of a single bird. I still remember a trip to Port Campbell National Park in Australia and watching the sky literally blacken with millions of mutton birds coming to roost for the night. Nothing prepared me for the awesome sight of those birds flying and veering in synchronized motion, converging as one on an islet.

Similarly, there is a difference between understanding the chemistry of a single water molecule and understanding the dynamics of billions of water molecules roiling in turbulence. Turbulence, in general, still remains an unsolved problem of physics. Whether it is a twirl of rising smoke or water rippling over smooth stones in a stream, classical physics cannot accurately predict the shape it will take or the force it will exert.

COMPUTER BUTTERFLIES THAT LED TO CHAOS

In cases such as these, scientists must study the whole. And the behavior of those wholes are not always predictable. Often they exhibit a tangled order. "Chaos" has become the shorthand name for a new research program within the scientific enterprise. James Gleick writes engagingly about all of the examples of complexity in the natural world that I have just mentioned and about scientists' attempts to understand them.

The name *chaos* might be misleading because it suggests that the physical world is in total confusion. In its Greek derivation, chaos meant a primeval emptiness, nothingness, sheer darkness. It did not mean disorder, which implies that there is something there to be disordered. By the seventeenth century, John Milton was still employing this original sense of the word when he depicted chaos in *Paradise Lost* as "the void profound / Of unessential Night . . . / Wide gaping, and with utter loss of being" (II, 438–40). The word *chaos*, as Gleick uses it, suggests rather that order and disorder exist side-by-side in nature. This is a revolutionary insight for science. We need only recall T. H. Huxley's case-closed pronouncement in the late nineteenth century: "The fundamental axiom of scientific thought is that there is not, never has been, and never will be, any disorder in nature" (1892, 247).

Well, it is a new century and the case has been reopened. Disorder does exist, and, Gleick explains, it can be modeled on a computer with even the simplest equations. Tiny differences in the input can result in vast differences in the output, producing apparent randomness and unexpected complexity. In the case of weather, this is playfully referred to as the Butterfly Effect. Gleick describes this as "the notion that a butterfly stirring the air today in Peking can transform storm systems next month in New York" (1987, 8).

There are now journals and scientific conferences devoted to the subject of chaos. Gleick chronicles the emergence of a new breed of scientists who are interested in the irregular, the nonperiodic, the nonlinear, the "messy" aspects of the natural world — including weather and the movement of animal populations, cardiac arrhythmias and nonlinear oscillators. Much of this research is done on computers with equations and mathematical models. One of these models suggests that chaos may be even more chaotic than we thought. In this particular model, there is not even a formula to predict an outcome until it actually occurs (Moore 1990).

Walter Freeman, a professor of neurobiology at the University of California at Berkeley, believes that chaos is a property of the brain itself and is evident in the tendency of the billions of neurons to shift from one complex pattern to another in response to sensory input. According to Freeman, chaos underlies the brain's ability to grow, to reorganize itself, and to flexibly and rapidly respond to the outside world. Writing about the physiology of perception, Freeman says that chaos "may be the chief property that makes the brain different from an artificial intelligence machine" (1991, 85).

Researchers are also looking for signs of chaos in the quantum world as well as in the solar system (Gutzwiller 1992; Hartley 1990). It is uncertain at this point where this research thrust will lead. The concept of chaos may transform our scientific thinking, as did relativity and quantum physics, or it may not. In any event, it has glimpsed an opening of truth that classical science refused to stare down.

SCIENTIFIC METHODS

Perhaps the most important discovery of twentieth-century science will be the extent of our ignorance. This includes, of course, our ignorance about science itself. Science is the most diverse of human pursuits. It takes in astrophysics, as well as the study of mountain gorillas in central Africa, or a virus such as AIDS. Fossils, pulsars, humpback whales, DNA, toxic waste, the human brain — they are all the province of science. At one time we thought that all this varied activity that we call science was brought together by a common approach — the scientific method. But the method often seemed too abstract and did not really describe what scientists actually do. One could not help but suspect that there needed to be a variety of methods to accommodate all this diverse investigation and experimentation.

Using the metaphor of a map, John Ziman explained why it is difficult to describe scientific method as a linear process (1978, 82–85). Like a map, a theory must configure data that is incomplete, approximate, and, at times, inaccurate. It must connect up information from many different directions, resolve any ambiguity, and, most importantly, be useful for further exploration.

Also concerned with what scientists do and not with what they say they do, Karl Popper rejected inductive reasoning as a basis for science and instead described the scientific method as one of conjecture and refutation. Popper's "criterion of demarcation" that separated the scientific from the nonscientific hypothesis was "falsifiability" not "verifiability" (1959, 40). Every scientific hypothesis must be capable of being tested and, therefore, capable of being proven false.

Popper, of course, would be uncomfortable with a vision of science such as Kuhn's. In his own way, Popper was shoring up the empirical method of the rationalists that had been under attack. In Popper's view of science, no scientific hypothesis can ever be proven true. No matter how many times the experimental results confirm the theory, you can never be sure that the next run of the experiment, the next observation, will not contradict the hypothesis. What you accumulate with each test is probability, not proof. There is reassurance in science, but, Peter Medawar explains,

> no matter how often the hypothesis is confirmed — no matter how many apples fall downwards instead of upwards — the hypothesis embodying the Newtonian gravitational scheme cannot be said to have been proved to be true. Any hypothesis is

still *sub judice* and may conceivably be supplanted by a different hypothesis later on. (1990, 99)

On the other hand, you can disprove a theory by finding just one observation that disagrees with its predictions. The basic requirements of scientific theory are that it explain current observations and that it predict new observations. Myth, for example, offers an explanation of the world, but it fails to be predictive.

THE CLOCK HAS STOPPED: UNCERTAINTY IN THE UNIVERSE

Predictability has been, therefore, a sine qua non of Western science. With the principle of uncertainty, however, even prediction becomes a dilemma for modern physics — the most "scientific" of all the sciences. According to Werner Heisenberg, it is impossible to simultaneously determine both the position and the momentum of a subatomic particle, such as an electron. He explained this principle with a type of mind-experiment just as Einstein had done with his theory of relativity. You must first imagine a microscope that uses gamma rays, which have a much shorter length than light rays. We can use this microscope to determine the position of the electron. In the process of finding its position, however, our gamma-ray microscope will bump the electron with a photon and change its direction and speed. So we cannot measure the electron's position and direction at the same time. When we focus on certain values, others change. In the act of measuring, the measurer changes what is measured.

Attacked by Einstein and many others, quantum mechanics was a new theory of mechanics based on the uncertainty principle. Accordingly, it cannot predict a single result for an observation. All it can do is predict a number of possible results (Hawking 1988, 55). "With the advent of quantum mechanics," Stephen Hawking discloses, "we have to come to recognize that events cannot be predicted with complete accuracy but that there is always a degree of uncertainty" (1988, 166).

One extraordinary consequence of the uncertainty principle is that reality is not "out there," posing, as it were, and waiting for the scientists to turn their instruments upon it. "The observer, and his apparatus," Lewis Thomas tells us, "*create* the reality to be observed. Without him, there are all sorts of possibilities for single particles, in all sorts of wave patterns. The reality to be studied by his instruments is not simply there; it is brought into existence by the laboratory" (1979, 71).

These remarks by Hawking and Thomas demonstrate just how much of a change there has been in our view of science from the time of Galileo and Newton and from the scientific determinism of Laplace. At the extremes of the physical sciences — when we are studying, for example, the very small or the very distant — interpretation of captured data replaces direct observation. Astronomers do not gaze expectantly at the night stars through telescopes any longer. They analyze the data from radio instruments or computer images.

In January 1992, for example, Aleksander Wolszczan and Dale Frail announced that they had discovered, with the help of the Arecibo radiotelescope, the first planets beyond our solar system. The planets were orbiting a millisecond pulsar, so named because they are neutron stars rotating hundreds of times a second. When Wolszczan and Frail determined that the radio signals from this pulsar varied just a few trillionths of a second from their expectations, they postulated the existence of two or more planets tugging on the pulsar as they orbited (1992, 146). Interestingly, though, it will take several more years of observation to determine if the planets actually do exist. A similar announcement of a planet discovery in 1991 proved to be an astronomical error — in every sense.

Is it observation when there is no human observer? Instruments do not observe. They measure changes in light, sound, temperature, and movement. Or they model these changes with programmed equations and streams of input. In any event, for billions of light years, the stars have not been in the "spots" where we are looking. Astrophysicists see things as they were, not as they are. Their instruments receive light and radio waves from "objects" that, in some cases, no longer exist. It is somewhat like getting the singing telegram fifty years after the messenger's funeral.

OUT OF SIGHT, OUT OF MIND: BLINDNESS AND BIAS

Today, scientific belief is grounded more and more on what cannot be observed by scientists, with or without instruments. In the 1970s and 1980s, for instance, there was interest among scientists in "string theory" to describe the strong forces between subatomic particles. This theory, though mathematical and precise, lacked any observational evidence at all — as much as the theory that the earth is supported by a neat stack of piggy-backed tortoises (Hawking 1988, 171). To a large extent, modern theoretical physics is, as Stephen Hawking describes it, a physics that is "all in the mind."

In other words, seeing is not believing anymore. Stephen Tyler writes in *The Unspeakable* about "the hegemony of the visual" in the Western tradition (1987, 150). We who have been educated in this tradition think of thinking and knowing as seeing. "I see what you're saying." "Can you shed some light on this situation?" "Look at it this way." "Let me show you what I have in mind." "Do I have to draw you a picture?" "You're missing the big picture." "The bottom line is . . ."

Tyler wants to say even more than this. For him, at least, science and logic belong "to a cultural tradition" and "cannot function as universals" (1987, 170). In predominantly oral cultures, hearing is believing. That is one reason Tyler disputes any claims of logic or science to universality:

> Except as ideology (that instrument of power), logic and science, those arts of the visual, have no more claim to universality than the traditions they pretend to dominate, for these traditions speak of other things and even remind us, in the words of a Bemba saying that: "The eye is the source of the lie." (1987, 170)

Certainly, propositional logic grew out of ancient Greek culture. Unlike Tyler, however, I think it is a completely separate question whether logic and the science it has fostered and promoted can increasingly function as global universals today. Both have already spread around the planet through scientific and academic institutions and practices. The whole issue of the universality of logic is muddied because we do not fully understand what we really mean by logic when it comes to actual reasoning in everyday situations, what the distinction is between logical reasoning and logical practice, and whether scientific reasoning is something apart from ordinary human reasoning. In a way, though, even we in the West have come to understand like the Bemba, a Bantu-speaking people in Zambia, that human sight is limited and easily fooled.

Tyler's own work, of course, carries forward the seminal scholarship of writers such as Eric Havelock, Walter Ong, and Frances Yates. Although their research has been in the humanities, their collective analyses of early written and printed texts lead up to the more far-reaching conclusion that our Western preoccupation with science and logic — and computer systems — emerge from our visual bias.

This visual bias persists in cognitive science today when we describe human memory in terms of encoding, storing, and retrieving. Memory becomes, in Tyler's expression, "a collection of loci" in which mental objects are stored — similar to the discrete, serial, addressable locations in a computer's memory (1987, 135).

The same bias is concealed in our discussions about knowledge representation for the computer. Tyler's concern is communication and the nature of meaning, not Artificial Intelligence or computers, but his insights are applicable nevertheless. "The idea of representation makes sense," Tyler points out, "only in a context of writing and visualization, of the spatialization of sound" (1987, 135). In an oral context, we might more properly speak of "evoking" reality rather than "representing" it (Tyler 1987, 207). Sound is called forth and emanates from an interior, while representations are fixed on surfaces. Tyler makes the observation that oral discourse is "the other as us" whereas "non-participatory textualization is alienation — 'not us'" (1987, 205).

GLOBAL VILLAGE

Computers, of course, assimilate to both oral and textual models. Though a highly formalized and text-based technology, computers and the electronic communication media that they sustain can draw persons into consciousness and participation in ways that print technology could not. Computers and television made possible the so-called town meetings during the 1992 U.S. presidential election campaign. The expression, in itself, showed advertising and political savy — with shades of Marshall McLuhan's "global village." But it also touched on the real power of these new technologies to increase our awareness of persons and events, to accelerate the flow of information, and to attenuate distances of space and time.

These distinctions between the way knowledge is perceived, organized, communicated, and understood in a hyper-literate yet hyper-electronic culture and in an oral or residually oral culture (generally fixated upon literacy as a national goal and priority) will be examined further in later chapters.

WRESTLING REALITY

As you move from particle physics to the social sciences, it becomes even more difficult to achieve scientific predictability because of the complexity of the object of scientific observation. In fact, it becomes more difficult to even define exactly what scientific observation is and what the object is that is being observed. We have recognized this for a long time now. Social reality is largely unpredictable and disorderly — a wriggling subject for science. This is not to deny that there are all sorts of group dynamics, posturings, and organization going on in our human societies and subsocieties, some of which the sociologists, psychologists, and anthropologists have observed and documented. It is to say, however, that social reality cannot be isolated or pinned down in a laboratory experiment.

This is just the subject I have before me now. When we attempt to talk about computerized knowledge systems and Third World environments, we no longer have real objects of study. What we have is an emerging technology, loosely defined and variously implemented, that will interact with complex social and political events and diverse ethical and cultural structures to transform human consciousness in ways that will not be fully open to analysis and that will always be dynamic and synergistic.

Chapter 2

The Downside of Definition

Everything should be made as simple as possible, but not simpler.

— Albert Einstein

A name is a prison, God is free.

— Nikos Kazantzakis

It is always one world that creates another.

— George Santayana

When put up against reality, all our distinctions and definitions begin to slip and slide. Concepts and definitions both facilitate and limit our understanding. As the tension becomes too great between the reality and the conceptualization, we cast a net for new terms, new definitions. Often this is just an exchange of one set of limitations for another. I will be working within just such limitations as well.

The Western mind typically approaches reality in terms of defined problems and solutions. This is true no matter if the reality is the atom, tropical rainforests, or Arab-Israeli hostilities. However, sooner or later — whether the subject is subatomic physics or the dynamics of international politics — we must move beyond abstraction and definition to a somewhat uncomfortable immersion in the world itself.

Certainly, the subject before us is wide-ranging. How do you begin to say something meaningful about such complex matters? You begin, I think, by slicing through some of the abstraction, by acknowledging the limitations of your definitions and your conceptualization of the problem.

AT THE FRINGES OF CONSCIOUSNESS: THE FOURTH WORLD

The concept and expression *Third World* is a prime example of reality spilling over the boundaries set by language. Although it may be in the process of gradual replacement by the term *the South*, it is still the expression most often used to identify the poorest nations of the earth.

All nations, of course, have their poor. The homeless who sleep in bus stations and beneath highway overpasses or seek out shelters at night in every city in the United States are part of our poor. Some even use the term *Fourth World* to describe undeveloped communities within First World nations. The poor, often along with minority ethnic groups, have been marginalized within these industrialized nations in much the same way as the Third World has been marginalized by the Western industrial nations as a whole. Native Americans are an obvious example of such marginalization in the United States. Speaking with such eloquence in the late nineteenth century of the white nation that moved inexorably west across the Great Plains, a Lakota survivor said: "They made us many promises, more than I can remember, but they kept but one; they promised to take our land, and they took it" (Eagle Walking Turtle 1991, 10).

In the last thirty years, we have been taught a new history of the American West, one in which honor and brotherhood are more often virtues of the Indians (a name some Native Americans now reject) than the white settlers and armies. Some of this is idealized and reminiscent of the "noble savage" literature of the nineteenth century, but much of it is accurate. Even today, the Iroquois have not accepted U.S. citizenship, but regard themselves as a proud and separate nation. This does not alter the fact that, for most of U.S. history, Native Americans have consistently been kept from power and relegated to the fringes of white consciousness. As elected chiefs assert self-determination of tribal rights within their own reservations — whether those rights apply to oil, gambling, land development, or the return of adopted Native American children — they are slowly regaining some economic and political power.

Likewise, while the black community of the United States has made progress since the Civil Rights movement of the 1960s, many continue to exist on the periphery of American society, living with substandard housing, education, health care, and justice. This, plainly, is the hands-off rhetoric of commission reports and government statements. There is more urgency in the television news film of a black man, Rodney King, being stomped and whipped by a ring of white policemen and of black retaliation and terrorism in the Los Angeles riots of summer 1992. There is more immediacy in the oratory of a political leader, such as Jesse Jackson, who speaks of black children growing up in a neighborhood of broken streets, broken schools, broken homes, broken dreams, and broken lives. But even these images and rhetoric disguise the reality. While we speak about the black "community," there is no homogeneous experience among its members. Today some black persons — educated in the best universities, highly paid professionals in their chosen fields, suburban homeowners — are removed from the experiences of "the brothers" and "the sisters" in the inner cities. It is difficult for them to really identify with the frustrations and anger there.

THE BUSINESS OF POVERTY

Clearly, economics creates marginalization just as much as race and ethnicity. With rising U.S. unemployment and loss of jobs due to an ongoing recession and to overseas competition in the opening years of the 1990s, even white middle-class Americans learned how easily and quickly it is possible to become invisible to the system — without income, insurance, or the social identity that employment confers.

At the same time, it would be a mistake to think that everyone who lives in the Third World is economically poor. Within these countries is great affluence concentrated in the hands of an elite — the politicians, the large landowners, the managers of local and multinational corporations, the Western-educated, the university graduates, the bureaucrats, the religious leaders. There has always been money to be made from the poor, whether you are talking about exploiting their labor, their weakness, their ignorance, their faith, or their dreams. Poverty is big business and good business. Historically, even the motives for U.S. foreign aid to poor countries have been more political and economic than humanitarian. In their book, *Food First*, Frances Lappé and Joseph Collins point out that, in addition to the official policy statements about "helping" the underdeveloped countries, these motives have also been to unload price-depressing surpluses, to create new markets, to negotiate access to strategic materials, such as oil and minerals, and to permit military intervention (1982, 364).

THIRD WORLD DEBT

The popular conception of foreign aid is that it is charity and that the people of the United States cannot afford to send money overseas when there is poverty within the streets of Detroit, St. Louis, Memphis, and thousands of other American cities and towns. What is not popularly known is that Western banks are making much more money on the developing world than the governments of the industrialized nations are "giving away" in foreign aid. Third World debt and interest payments in 1988 alone amounted to $178 billion. This is three times the amount of aid that the industrialized countries sent to the Third World (UNICEF 1990, 1).

According to a White House document issued in 1990, the "aggregate Third World debt is over $1 trillion, and debtor nations need some $70 billion just to meet annual interest payments" (*National Security* 1990, 21). Inevitably, much of this debt is currently uncollectible and interest on the debt continues to accumulate. Increasingly, there is discussion about the need for Western banks to write off this Third World debt. The motives are more economic than humanitarian. Banks would alleviate part of the debt in order to stimulate the market for Western manufacturers and allow them to sell more products overseas. This would also allow the banks to make new loans.

As a whole, the governments of the developing countries now devote half of their total annual expenditures to paying their national debt and to buying military hardware from Western companies (UNICEF 1990, 1). Clearly, these governments must also accept some responsibility for the crises within their nations: for poor choices of development projects, for corruption that siphoned off foreign aid, for military buildups used more often to suppress their own people than to defend them against the aggression of an outside enemy, for institutionalized injustice that kept land and privilege in the hands of a few, for civil wars and ethnic purges. Denis Goulet, however, points out that when power is abused, we cannot simply blame the "personal deficiencies of rulers" because "the problem is structural: society is so organized that only the representatives of certain interests enjoy access to the wealth, culture, contracts, information, and influence without which decisions cannot be made" (1971, 336).

In any event, it is too simplistic to describe Third World nations as innocent victims and the West as greedy oppressors. There have certainly been victims and oppressors and very real suffering in the colonial and postcolonial relationships of rich and poor nations. The crushing poverty that exists in the Third World, however, has a tangle of causes.

NATIONS AT RISK

No matter what the causes, the dimensions of Third World poverty are immense. According to statistics released by the U.S. Department of Commerce, 75 percent of the earth's population live in developing countries (Jamison 1989, 1). That number, also according to projections, will rise to 80 percent within the next thirty years (Jamison 1989, 1). For the most part, these countries are in Africa, Latin America, and the southern portions of Asia. Although many of these countries are in the northern hemisphere, they are still generally to the south of the industrialized nations of the First World. That is the reason some writers from the Third World prefer to use the term *the South.*

The crises within the Third World have become almost familiar dramas to us. The Western world's conscience is jolted with the evening news that up to three million famine victims will die in Ethiopia. Rock concerts and TV specials are quickly organized to raise food aid. A few years pass, and it is starving Somalians that have become the "human interest" story — victims of drought, civil war, and the looters who carry off the airlifted sacks of grain and boxes of medical supplies for the black market. The average viewer of these media events, however, is unaware that it is often day-to-day living that has become the crisis for the poor of the Third World. Fifteen million children below the age of five die silently and unnoticed by the news cameras every year, generally from diseases that can be prevented or treated inexpensively (World Health Organization 1987, 11). These fifteen million deaths are just one of the statistics that become part of our abstraction of poverty in this other world, this Third World.

THREE WORLDS — ONE EARTH

Even the concept of a Third World is an abstraction from our concept of the world. Today, most of us mysteriously carry within our brains some image of the world as a whole. Many of us have seen those extraordinary photographs from space of our planet suspended in cosmic darkness. As human persons — no matter what our color, our nationality, our religious or political convictions — we share this planet. There are no national boundaries from space. There is no line cutting the globe into First World and Third World, North and South. Spinning in space, earth has no eastern countries or western countries. This image of a shared home and a shared destiny would have been inconceivable before our generation of space exploration.

We have come to this image of global cohesiveness only lately and hesitantly and still incompletely. But the idea of "the world" — meaning both planet and planetary community — continually reasserts itself in our language. "The world is threatened by the depletion of the rainforests." "The world anxiously awaited news of Nelson Mandela's release from his prison on Robbins Island."

The reductionist tendency, however, is ever present, and we attempt to understand that world by separating it into like parts again. So we separate the Western democracies from the Eastern socialist states, and the industrialized, affluent First World from the agrarian, poor Third World. Whether intended or not, Third World is a pejorative term that carries with it, for many English speakers, connotations of third rate and third class. This perception is often shared by those who live in the Third World as well, particularly those who have been educated at universities and sensitized to labels and images.

DEVELOPING COUNTRIES

Even the term *developing countries* implies that these nations are not full-fledged and mature as the developed countries. The implication is also that the developed countries have achieved or arrived at the goal of the development process. There is an absurdity in all of this. China and India, for example, are the two largest countries in the world. Their combined population is almost two billion people (Jamison, Johnson, & Engels 1987, 40). This is about 40 percent of the earth's total population of five billion people (Jamison, Johnson, & Engels 1987, 1). Both India and China have civilizations that stretch back thousands of years. Their civilizations have made great contributions to knowledge and art and religion. Among other things, China is responsible for the invention of paper and the compass — both of which literally created world history. Gunpowder, which forever changed world history, also originated in China. One does not even have to be a scholar of Eastern studies to be aware of some of the archaeological treasures from both China and India. Many of us have seen articles in magazines, such as *The National Geographic*, with photos of these treasures. We have marveled at the 7,500 life-size clay soldiers — their robes and armor and horses all meticulously detailed and individualized — that were buried twenty-two centuries ago in the grave of China's first emperor, Qin Shi Huangdi

(Danforth 1982, 173). We have seen countless pictures of the beautiful Taj Mahal of India, one of the manmade wonders of the world. These two civilizations have given us teachers, such as Buddha, Confucius, and Gandhi, who have enhanced our vision of our own humanity.

Yet, China and India are considered developing nations while the United States, with its 200-year history, is a developed nation. The message is clearly that development is primarily an economic and political concept.

There is unintended irony of another sort as well in the term *developing nations*. In 1989, for every 100 persons added to the earth's population, 93 were in developing nations (Jamison 1989, 1). Third World debt continues to escalate, particularly in Latin America. Latin America's debt is now four times its total annual exports (UNICEF 1990, 12). In the 1980s, the number of children out of school in the developing world increased another ten million from the previous decade (UNICEF 1990, 9). There is definite development taking place in these nations — developing population, developing debt, developing unemployment.

None of the current terminology — Third World, developing world, the South — used to describe the discrepancy between the affluent and the poor nations of the earth is satisfactory. We are looking for an inoffensive way to describe poverty and injustice — and there is none.

THE VIOLENCE OF LANGUAGE

The political corollary of being a Third World nation is that you are frequently excluded from international decisions. Often decisions are forced upon you. These include economic decisions, such as the price a nation can get for its export of coffee or the price it will have to pay for its import of oil, as well as political decisions, such as border extents or, in the case of Palestine, the dissolution of an entire nation. Even the most powerful nations are frequently influenced by the international market and by international cooperatives, such as OPEC, but they have greater resources to withstand the pressures.

More often than not, the Third World is simply "lumped together." In U.S. history, we did this as well with the indigenous people — comprising several nations and hundreds of cultural groups, speaking utterly diverse tongues from at least seven language families — that became in our consciousness simply "Indians" rather than the Cherokee nation or the Navajo nation. In a White House document on national security that was issued in 1990, the Third World is generalized in this way and described as a potential threat to the United States:

> In a new era, we foresee that our military power will remain an essential underpinning of the global balance, but less prominently and in different ways. We see that the more likely demands for the use of our military forces may not involve the Soviet Union and may be in the Third World, where new capabilities and approaches may be required. We see that we must look to our economic well-being as the foundation of our long-term strength. (*National Security* 1990, 15)

The subtle implication appears to be that "our economic well-being" will be protected with "our military power."

The same White House document also alludes to the fact that new weapons and strategies will be developed for war in the Third World:

> Defense investment faces a dual challenge: to maintain sufficient forces to deter general war while also giving us forces that are well suited for the more likely contingencies of the Third World. Many defense programs contribute significantly in both environments but, where necessary, we will develop the weaponry and force structure needed for the special demands of the Third World even if it means that some forces are less optimal for a conflict on the European central front. (1990, 24)

In this official policy statement, the United States government reserves the right to act on its own in the Third World in order to protect its interests: "To the degree possible, we will support allied and friendly efforts rather than introduce U.S. forces. Nonetheless, we must retain the capability to act either in concert with our allies or, if necessary, unilaterally where our vital interests are threatened" (1990, 26).

Unfortunately, violent words often precede violent acts. The rhetoric in this national security statement became reality in January 1991 with the opening missile strikes of the Persian Gulf War. It is important to recall some of the details of that war because it represented a crucial decision by the United States and the United Nations Security Council to unleash the overwhelmingly superior technology of a Western alliance against an intractable government leader of a Third World nation, his armies, and his people. The morality of that action, I believe, will be debated for many years to come.

A DEADLY STORM

During a televised speech on July 17, 1990, Saddam Hussein, the president of Iraq, threatened to use force against Kuwait and the United Arab Emirates in order to stop them from driving oil prices down, a plot that he said was "inspired by America" (*Facts* 27 July 1990, 549). In debt from his long war with Iran, Hussein also had his eye on Kuwait's reserve of hard currency.

On August 2, 1990, one week after a controversial meeting between Hussein and U.S. Ambassador April Glaspie, the Iraqi army invaded Kuwait in a move toward annexation. Economic sanctions against Iraq were immediately set in place. In the ensuing months, a "coalition force" was assembled in the Persian Gulf. The troop deployments for the coalition included approximately 425,000 U.S., 25,000 British, 15,000 French, and 104,000 Arab soldiers (*Facts* 17 January 1991, 29).

There was ongoing debate about the effectiveness of sanctions. As a diplomatic stalemate developed, there was also increased speculation that military commanders would not be able to sustain the morale of 600,000 troops on hold in the Persian Gulf and that the Bush administration would not be able to keep the political support of the American people. The long and mired history of the U.S. involvement in the Vietnam War, the U.S. military's failure to achieve a victory there, and the war's increasing unpopularity with the American people were replayed by the political commentators. By November, the UN Security Council, under diplomatic pressure from the United States, adopted Resolution 678, authorizing "all necessary means" to remove Iraq from Kuwait (*Facts* 17 January 1991, 25).

President George Bush issued a further Security Council ultimatum that Iraq was to withdraw from Kuwait by January 15. With the deadline less than seventeen hours past, the United States initiated air and missile attacks on Iraq and began a campaign that was labeled Desert Storm (*Facts* 17 January 1991, 27). During the first fourteen hours of the war, U.S. planes flew 1,000 bombing missions over Iraq and U.S. ships launched 100 Tomahawk cruise missiles (*Facts* 17 January 1991, 27). The daily pounding of bombs and missiles did not stop until the cease-fire on February 28, 1991.

In the closing hours of the war, thousands of Iraqi soldiers who were fleeing from Kuwait City to the Iraqi port of Basra were trapped in a mile-long traffic jam. The Iraqi soldiers had commandeered tanks, private cars, buses, police cars, ambulances, fire trucks, and any available vehicle to make their escape. U.S. aircraft bombed and strafed the road that came to be known as "Ambush Alley." In viewing the carnage, one British soldier was reported to have said, "No human being should be allowed to do this to another human being. . . . They didn't stand a chance" (*St. Louis* 9 March 1991, B1). A Marine officer referred to it as a "duck shoot," and a British officer, viewing the carnage, said: "I guess the strategy was to destroy them once and for all, but it is horrendous.

There's not much art in this. We obviously just overwhelmed them" (*St. Louis* 9 March 1991, B1).

The months following the end of the war led to further revelations by the U.S. Defense Department. In response to a front-page article in a New York paper, the Defense Department admitted that scores of Iraqi soldiers, entrenched along the border between Iraq and Saudi Arabia, had been buried alive during the opening hours of the ground assault. U.S. tanks, equipped with plows, pushed tons of desert sand and earth into the Iraqi trenches. A spokesman for the Defense Department defended the tactic by arguing that it prevented further U.S. casualties and that it was not in violation of the Geneva Convention, which governs the conduct of war. He added: "There is no nice way to kill somebody in war" (*Facts* 19 September 1991, 686).

The war apparently played out just as U.S. military leaders had predicted. In a press briefing one week after the war began, General Colin Powell, chairman of the Joint Chiefs of Staff, laid out the strategy for the war with chilling bluntness: "Our strategy for going after this army is very, very simple. First, we are going to cut if off, and then we are going to kill it" (*St. Louis* 9 April 1991, B1). A Pentagon planner referred to the war as "almost an industrial operation" since the Iraqi army was obviously mismatched for a battle with the United States' high-tech weapons (*St. Louis* 9 April 1991, B1).

At the start of the Desert Storm campaign, Iraq reportedly had a conscripted army of one million men — and boys. Numbers did not matter, however, because Iraq's troops never engaged the coalition forces until the closing four days of the war. This was an air war of unprecedented ferocity.

Throughout the war, American television viewers were "educated" about the effectiveness of the U.S. military's high-tech weapons against this Third World army. Film footage of F-117 Stealth fighters locking onto their targets with lasers and surgically destroying them were daily fare.

ENFORCING A NEW WORLD ORDER

Only after the war, did more accurate reports of what the Pentagon labeled "collateral damage" begin to surface. General Merrill McPeak, the U.S. Air Force chief of staff, relayed on March 15 that 88,500 tons of U.S. munitions had been dropped on Iraq and occupied Kuwait in the course of the war. Only 7 percent of these had been laser-guided "smart" weapons with an accuracy of 90 percent. The other 93 percent or 81,980 tons had been unguided bombs with an accuracy of only 25 percent (*Facts* 28 March 1991, 215). Their collateral damage had left 72,000 civilians homeless. There was no estimate of civilian casualties, however, a UN team visited Iraq from March 10 to 17 and filed a report that said the bombing campaign led by the United States had

> wrought near-apocalyptic results upon the infrastructure of what had been until January 1991 a rather highly urbanized and mechanized society. . . . Now most means of modern life support have been destroyed or rendered tenuous. Iraq has . . . been relegated to a pre-industrial age, with all the disabilities of post-industrial dependency on an intensive use of energy and technology. (*Facts* 28 March 1991, 214)

President Bush, while referring constantly to a "new world order," said that one message from all the American firepower was that "what we say goes." After the war, Bush indicated that he might have sent American troops to the region even if the United Nations had not authorized aggression against Iraq: "I might have said 'to hell with them. It's right and wrong; it's good and evil. He's evil, our cause is right,' and, without the United Nations, sent a considerable force to help" (*St. Louis* 10 March 1991, 9A). He also alluded to the popular disillusionment and criticism following the Vietnam War and said, "By God, we've kicked the Vietnam syndrome once and for all" (*Facts* 7 March 1991, 157).

The Persian Gulf War was supposed to have restored America's confidence in its status as a superpower. The missiles had scarcely stopped flying before American television carried ads for videotapes so that we could relive the excitement of our top-guns for only $29.95. Towns and cities across the United States organized parades and church bells rang out in celebration for the returning troops. It seems the only way to end war is to make it unprofitable — for the politicians, for the industrialists, for the media, and for the patriots.

General Norman Schwarzkopf, the commander of the coalition forces in the Persian Gulf, even indicated at the end of the war that he had not been in favor of the cease fire. Instead, he wanted to "annihilate" Iraq's armies as Hannibal had annihilated the Roman army at the Battle of Cannae in 216 BCE. In a televised interview, Schwarzkopf said: "Frankly, my recommendation had been, you know, continue the march. I mean, we had them in a rout and we could have continued to reap great destruction on them. We could have completely closed the door and made it a battle of annihilation" (*St. Louis* 27 March 1991, 10A).

The ruthlessness of Saddam Hussein was, in fact, not the issue of debate — either before, during, or after the Persian Gulf War. That had already been well-established in his treatment of the Kurds, a large ethnic population without a country. Although they are not Arabs, Persians, or Turks, one segment of their population lives in the mountains where the borders of Iraq, Iran, Turkey, and Syria come together. In 1988, Hussein used poisonous gas against the Kurds and killed 5,000 villagers, including women and children. When films of these atrocities were shown on American television, there was no reported censure against Hussein from the U.S. government. Even up until the day before Iraq's invasion of Kuwait, the Bush administration had approved the sale of technology to Iraq and sought closer cooperation with Hussein's government (*Facts* 28 March 1991, 213). During and after the conflict, however, the administration expressed concern that Iraq was developing nuclear weapons and that Hussein would not stop at using them.

Atrocities of the occupying Iraqi army against the Kuwaiti people were documented and reported during the war. In addition, Saddam Hussein literally released a tide of destruction and created ecological disaster when he ordered millions of gallons of oil to be spilled in the Gulf. Between the cease-fire on February 28 and the formal end of the war on April 11, 1991, Hussein again turned his helicopters and military weapons against Kurdish and Shiite Muslim insurgents within his own country who had been given U.S. encouragement to rebel. Thousands were killed and, reportedly, 1,500,000 refugees were forced to flee Iraq into Iran and Turkey (*St. Louis* 14 April 1991, 1). Following the end of war, Hussein, still in power, continued to carry out virtual genocide against the Kurds in the north and the Shiite Muslims in the south. During this time, the United States limited its involvement to airlifting food and medical supplies, enforcing sanctions, creating no-fly zones for Iraqi aircraft in the north and south of Iraq, and threatening the government of Iraq if it did not comply with UN demands for weapons inspection. On January 13, 1993, these threats erupted in renewed missile strikes by U.S. and allied planes on military targets near Baghdad as Saddam Hussein continued to violate the peace agreement by crossing into Kuwait to retrieve Silkworm missiles, refusing to allow UN inspection, and ordering Iraqi antiaircraft to fire on U.S. planes penetrating the no-fly zone.

BODY COUNT

The issue of debate, therefore, was not the brutality or defiance of Saddam Hussein. It was the U.S. military response to Hussein's posturings — and the ferocity of that response. Desert Storm was a war of high technology in which the casualty lists pointed tragically and unforgettably to the inequities of nations — 148 U.S. combat deaths to perhaps as many as one-quarter million Iraqi soldiers and civilians. The Saudi Arabian ambassador to the United States gave the estimate for Iraqi dead at 85,000–100,000 soldiers while U.S. military analysts initially released a figure of 25,000–50,000 soldiers (*Facts* 28 February 1991, 125). Later, General Norman Schwarzkopf presented the Bush administration and congressional officials with estimates of at least 100,000 Iraqi soldiers killed (*Facts* 28 March 1991, 215). Others outside the U.S. military and government, however, have judged the actual count of the Iraqis who died directly from the bombs and bullets, as well as from the environmental devastation that resulted in diseases; water pollution; and lack of sanitation, food, medical supplies, and medical care, between 200,000 to one-quarter million people (Clark 1992, 206–09). Total allied deaths, including the 148 U.S. soldiers, were reported as 223 (*Facts* 25 July 1991, 559). On August 13, 1991, the U.S. Defense Department admitted that almost 24 percent of the U.S. casualties, or the deaths of 35 American soldiers, were the result of our own friendly fire (*Facts* 15 August 1991, 609). In addition, 72 of the 467 Americans wounded in combat were victims of our own weapons (*Facts* 15 August 1991, 610).

Colonel Ralph Hayles, who was disclosed as one of the U.S. battalion commanders responsible for mistakenly firing on a U.S. ground vehicle, said in an interview that the deaths only became a horror when it was confirmed that U.S. solders had died ("Friendly Fire" 1991). When the vehicle was assumed to be Iraqi, the missiles were launched with a "toast" — "This Bud's for you!" Throughout history, societies have denied humanity to their enemies. With the Greeks, it was the barbarians. Both Hayles's admission and the repeated reference of U.S. politicians to the "minimal" casualties in the Persian Gulf War even though thousands of Iraqis had been killed are indications that we still fail to see our common humanity despite racial and cultural differences.

The Persian Gulf War was a striking real-world example of international politics, in which the United States, assuming a leading role for the United Nations, was pitted against an oil-rich, heavily armed but technologically less-developed nation — in which Western attitudes dueled with Eastern. The roots of the conflict go back, of course, beyond the economic and political tensions of the Middle East in the 1980s and 1990s. The Middle East nations were created by Britain and France at the Uqair conference in 1922. Border lines were drawn in the sand without regard to the ethnic or religious communities who lived in the region for thousands of years. The strategy was to satisfy the colonial interests for oil and to prevent any of the new nations in the Middle East from emerging as a power that could threaten those interests. The creation of the state of Israel in 1948 caused additional resentments and tensions.

CHALLENGES FOR THE SOUTH

We have to keep realities like these before us or our discussions about technology transfer and Third World environments become hopelessly abstract. It is not a matter of interjecting politics into what should be a purely technical consideration. Technology transfer and development are inherently political issues.

The Persian Gulf War also graphically demonstrates why the poor nations of the world are both enamored of our technology and justifiably sensitive to the military power of the West, particularly of the United States. There is tragic irony in the fact that the five permanent members of the United Nations Security Council — the United States, the former Soviet Union, France, China, and Britain — are also both the world's biggest arms sellers and the owners of the biggest nuclear arsenals (*St. Louis* 3 July 1991, 10A). Over the next decade, the Strategic Arms Reduction Treaty (START I and II) calls for a gradual reduction of the arsenals only in the United States and in the republics of the former Soviet Union in which warheads are based — Russia, the Ukraine, Kazakhstan, and Belarus — but still leaves the planet burdened with nuclear proliferation and destruction.

One document, *The Challenge to the South*, a report of the South Commission that consists of representatives from all the Third World regions, speaks to this very point of Western militarism:

> The widening disparities between South and North are attributable not merely to differences in economic progress, but also to an enlargement of the North's power vis-à-vis the rest of the world. The leading countries of the North now more readily use that power in pursuit of their objectives. The "gunboat" diplomacy of the nineteenth century still has its economic and political counterpart in the closing years of the twentieth. The fate of the South is increasingly dictated by the perceptions and policies of governments in the North, of the multinational institutions which a few of those governments control, and of the network of private institutions that are increasingly prominent. (1990, 3)

This is not to deny that the Third World has its tyrants and repressive governments, and its own militarism and aggressive rhetoric. It has. In recent years, civil wars have killed hundreds of thousands of people in places such as Lebanon, Uganda, and El Salvador.

TYRANNY OF MADNESS

It is a mistake, therefore, to speak as though all Western technology and culture will corrupt the Third World nations. George Bush, for instance, with UN approval and with support of the principles of international law, sent U.S. troops into Somalia in December 1992. The government of dictator Mohamed Siad Barre had collapsed in January 1991. Former military officers and their followers, armed with weapons once supplied by the United States and the former Soviet Union, turned the main towns of Mogadishu, Bardera, and Baidoa into a tyranny of madness. Bullets were used to gain control of the only thing left to fight over — food. Looters stole food and medical supplies that had been collected by international relief agencies for the million Somalis threatened with starvation. The independent state, created in 1960, ceased to exist.

In this chaos, questions about intervention and sovereignty were largely moot, and a First World military force was marshaled for a humanitarian mission in a Third World nation. If there were reservations about Operation Restore Hope, they generally focused on what would happen after the troops were withdrawn, how a government could be created, and how that government could successfully unite northern Somalia, a former British colony, with southern Somalia, which had once been governed by Italy. Some Somalis suggested re-empowering the institution of the elders, the indigenous form of government that helped to maintain relative peace in the outlying areas of the country, although it is difficult for Westerners to understand how this traditional institution could function as a modern state.

Injustice and prejudice and ignorance exist within Third World societies just as they do within our own. Even though showing respect for other cultures, for example, I cannot condone the fact that in African, and in Far Eastern and Middle Eastern countries millions of women have been subjected to genital mutilation as a safeguard against marital infidelity and disputed patrimony. There are things, therefore, that need to be changed in these societies as there are in our own.

REVOLUTIONARY WORDS

Certainly, some Western ideals have transcultural value. It is not by accident that some of the Latin American governments have constitutions that adopt the wording of the United States Constitution and the Declaration of Independence. These last two are extraordinary documents. For the first time in human history, a nation came into existence when a group of human persons, with common intent and consent, gathered together and affixed their signatures to a piece of paper which was copied and circulated in print. These documents embodied the new validity, authority, and power of the printed word and textualism. The ideals that the documents express — democracy, equal justice, peace — are the best of human aspirations. However, they are ideals that have not yet been fully realized either within our nation or in our relationships with other nations.

THE TRAP OF GRAND THEORY

In considering the introduction of a computer technology such as expert systems, it is important to realize that Third World nations will vary widely in their degree of indigenous technological preparation and their potential for participation in any system development process. Some of these nations, in spite of being Third World, will have computer industries in place. These industries include manufacturing, maintenance, and support for computer hardware and software. Other nations lack all of these.

We also cannot generalize about the leaders of the Third World and assume that all of them are clearly interested in the issues of economic growth, equity, technology transfer, and all the human priorities that analysts and commentators feel are essential to development. Some are not. Some are Western-educated and are anxious to accumulate Western technology as rapidly as possible, either to enhance their own positions or because they have not questioned the general Western assumption that development and well-being are advanced by each technological innovation.

Above all, it would be a mistake to think that technology itself (even the latest and the greatest) will be a total solution to Third World problems. It has not created an equitable and just society in the West either. The dimensions of Third World poverty are too great for quick fixes. Clean water, sewage, food, shelter, jobs, and medical treatment are higher priorities than microchips and software. In some cases, the technology of the oxen would better serve the poor themselves than the finest expert systems for agricultural planning.

Such diversity has clear implications for modeling and theorizing about expert systems in the Third World. It will be impossible and futile to attempt Grand Theory. That is not the intent or the design of the remaining chapters. At best, I can offer an analysis of the technology and some guidelines for its use that follow from that analysis. These guidelines will just be a subset of some of the larger questions that developing nations must address about computer technology in general, even before expert systems are considered:

1. What can computer technology effectively be used for within the current national setting and in light of national goals?
2. What effect will the introduction of computer technology have upon the social, political, and cultural structures of the nation?
3. Will the introduction of computer technology advance social and economic equity? How can this be measured?
4. How can the computer technology that is currently in the Western marketplace be effectively evaluated by developing nations? What are the criteria? What are the resources of information?
5. What are the educational and psychological prerequisites for introducing and using this technology?
6. How will the technology be managed? Who will manage it? Will maintenance of the technology require foreign experts and consultants?
7. Will the introduction of computer technology incur further dependency on the West for hardware, software, and ongoing training and maintenance?
8. What ethical issues are involved in introducing this technology?

One of the challenges for the leaders of the developing nations today and for the NGOs (nongovernmental organizations) that have become active in development work is to do this uncomfortable self-examination as a preliminary requirement for any strategic planning that will involve expert system technology.

Some will regard the topic I have introduced in this chapter as falling generally within the area of technology assessment. In a sense, it is. However, again we are faced with a problem of definition. A hydroelectric dam, a process for in vitro fertilization, a laser for microsurgery — these are all forms of technology and products of human knowledge. In the case of biotechnology, the products are more than just technology. They are life forms or processes that can alter life forms and even human genetic structures.

Computerized knowledge systems, however, are not simply products or artifacts of human knowledge. Expert systems incorporate human knowledge and, on some still-rudimentary level, the human reasoning process to uncover knowledge. They express our values, our definition of problems, our solutions to problems. Therefore, strategies to transport this technology from Western nations to the Third World must address issues that have not been previously addressed.

The computer revolution that writers like to tout is part of technological change. Technological change is not simply a matter of causes and effects. In the past, however, we have gone about the task of technology assessment as though it were. We busied ourselves discussing the impact of this technology or that one and measuring how these impacts had rearranged our human environment after the fact. We generally did not ask whether the design and choice of a technology compels a particular way of life, a particular structuring of social roles, personal relationships, and attitudes. Those are the questions, however, which must be asked. We must constantly move toward new understandings of ourselves and our technologies. The failures of development theory and Artificial Intelligence so far have shown us that many of our preconceptions about development, expertise, knowledge, and logical reasoning have been misconceptions.

Chapter 3

Human Passages: Toward a New Theory of Development

A complete revolution, perhaps man's last and greatest one, has become necessary. It will be more long-drawn-out than others of the past, because it must be universal. It is both a political and an economic revolution; but it can come to its fulfillment only if it is primarily a spiritual revolution — the expression of consideration for others because of their actual or potential value as human beings.

— Louis-Joseph Lebret

Our scientific power has outrun our spiritual power. We have guided missiles and misguided men.

— Martin Luther King, Jr.

Development is something that is going on continually — within individuals, cultures, institutions, and species. Some of this development is beyond our awareness. Some is beyond either our ability or our freedom to direct it. We are not really certain about all the dynamics that bring about development. We are not even in agreement among ourselves about what development really is. In a sense, we cannot speak about development as though it were simply some process separate from the individuals, the cultures, the institutions, and the species that are actually changing — or separate from the whole social and ecological context in which each of these exists.

It is well to remind ourselves of all this when we choose to talk or write about development theory. Bjorn Hettne, among others, counsels that "development is more complicated than its many doctrines" (1990,

251). He stresses that development studies, because of their complexity, must be interdisciplinary. "There can be" he adds, "no fixed and final definition of development, only suggestions of what development should imply in particular contexts" (1990, 2). Certainly, the concept of "three worlds" is becoming less useful and less acceptable, while development theory is becoming recognized as a global concern for rich and poor nations alike. This shift has come about, according to Hettne, because of two processes: self-criticism in the West and the "indigenization of development thinking in the Third World" (1990, 5).

My brief remarks in this chapter on development theory are not meant to be an intellectual history of the field. There are more substantive surveys for that. If our topic is the relevance of expert systems to Third World development, however, we cannot avoid some clarification of what development is and some explanation of how expert systems contribute to it.

Why are some nations rich and some poor? Why are some populations technologically advanced, with computer networks, laser tools, and space satellites, while others still farm the earth with hoes and oxen? Why are some states politically stable and others continually struggling with civil war and political coups? Why are some national leaders powerbrokers and decision makers on the world scene while others are marginal, invisible, or even political and economic pawns?

Until quite recently, the assumption of economic and political development theory was that it was better to be rich and powerful than poor and weak. The development theorists analyzed the causes and the cures for underdevelopment, while the development workers, from the Agency for International Development (AID) officers to Peace Corps volunteers, were expected to bring about the actual transformation.

GETTING OFF THE PARADIGM: CAPITALISM AND SOCIALISM

The development theories that have dominated development strategy for the last fifty years pitted two ideologies against each other: capitalism and socialism. With the collapse of the socialist state in the former Soviet Union, these very terms — capitalism and socialism — seem more tenuous than ever. It is easy to argue that capitalism has become passé even in the United States where the government manipulates economic and monetary policy. For at least the last 100 years, since the publication of the English version of Karl Marx's *Das Kapital* in 1886, we have been conceptually locked into these two ideologies and find it difficult to do more than offer variations on an old theme.

The capitalist paradigm for development primarily included growth theory and modernization theory. The exemplar of growth theory, of course, was W. Arthur Lewis's *The Theory of Economic Growth*, and the chief text of modernization theory was W. W. Rostow's *The Stages of Economic Growth*. While there are variations on each of these theories, basically they all assumed that development was a matter of accepting and implementing Western forms of political and economic organization, of creating free markets and economies open to foreign investment and trade, and of internalizing "modern" attitudes.

It was a process, as William Adams points out, that was meant to recreate the Western world: "industrialized, urbanized, democratic, and capitalist" (1990, 4). Denis Goulet suggests that it was also an ambiguous process that was "often depicted as the crucible through which all societies must pass and, if successful, emerge purified: modern, affluent, and efficient" (1971, 13). While the process might be ambiguous, the measure of development was not. It was all in the numbers: rising GNP and increased per capita income.

The capitalist paradigm essentially reduced the value of human existence to economic productivity and postulated that development decisions were economic, technical, and political. It was, according to Adams, "the imposition of the established world order on the newly independent periphery" (1990, 5).

Over twenty years ago, Ivan Illich saw that "the plows of the rich can do as much harm as their swords" (1973, 357). He had only scathing criticism for the type of development that manipulates human needs into consumer wants and packaged solutions, all through intensive marketing. For Illich, these consumer wants included everything from Coca-Cola to compulsory schooling. He recognized that those who define development goals and policies generally do so in familiar ways which "satisfy their own needs, and which permit them to work through the institutions over which they have power or control" (1973, 368). Illich saw only a setup for failure in this formula for development. His judgment was sound, because the failures of high-priced, high-tech development projects in the poor nations of the world are readily documented.

While the advocates of economic-growth and modernization theories monitored closely the rise and fall of GNP and prepared comparative national scorecards that showed the economic winners and losers, they did not look at the people themselves and whether the daily lives of the poor had, in fact, changed. In many cases the gain in GNP went into the pockets of government bureaucrats, international corporations, or wealthy local businessmen. Sometimes it went to refinance a national debt or to build a military infrastructure.

Another offshoot, growth with equity theory, did redirect attention to the distribution of wealth and highlighted the need to look at unemployment levels, income distribution, and development services to the rural as well as the urban areas. The focus, however, was still economic.

The socialist paradigm, on the other hand, placed greater emphasis on the causes of underdevelopment rather than on solutions. The solutions were often vaguely phrased in terms of increased political activity on the part of the socialist opposition or even of revolution. The prime expression of this socialist paradigm was dependency theory. Writers such as Fernando Cardoso, Immanuel Wallerstein, Andre Frank, and Samir Amin blamed underdevelopment on a global capitalist system. Reacting to the eurocentric bias of development theory, the dependency theorists, or *dependencistas*, attempted a unification of two of the important movements of the twentieth century: Marxism and Southern nationalism (Smith 1986, 26).

Drawing largely on the historical experiences of Latin America, the *dependencistas* sought to explain why political independence in the South did not lead to economic independence. Their analyses divided the world into two zones. At the core were the Western industrialized nations: the profit takers, the rich and the powerful. At the periphery were the profit makers, the poor and dependent. Of course, if we choose to use this image of core and periphery, we need to be aware that the core has shifted during history. China, India, and the Islamic countries of the Middle East were once the great centers of civilization and commerce, while Europe was the periphery.

William McNeill argued convincingly in the 1960s that, by the fifteenth century, the major civilizations were on a par and that the "rise of the West" could not have been predicted. If anything, the civilizations of the Middle East and of India would have been the safer bet to dominate the world economically, politically, and culturally. Being at the core or at the periphery, though, has as much to do with mental attitudes as with economic, political, and technological power. These are attitudes that define reality and normality for one's own culture. Those at the core not only exclude those at the periphery; they often fail to see them at all. This has certainly been true of the ethnocentrism of the Western tradition. When Aristotle and Kant and Descartes wrote about the mind or the rights of man, they were referring to European man, not to the Maya or the Guarani or the Zulu — and certainly not to women. For most of Western history, this ethnocentrism was perhaps understandable — but it can no longer be. Those at the periphery, on the other hand, cannot afford to ignore those at the core.

We are just beginning to understand, as well, that history depends upon who is telling the story. In 1992, there was a debate in the media about whether we should celebrate the 500th anniversary of Columbus's first voyage to Central America. For the Indians of North and South America, Columbus's "discovery" meant enslavement, destruction, and, in some cases, extermination of whole Indian nations. Even though some mestizos now identify more with their European ancestors than with their Amerindian ones, it is insensitive to say that the survivors of this confrontation with Europe and their offspring were ultimately better off because they were given the benefits of Western institutions, technology, and Christianity. In fact, to coincide with the Columbus anniversary, the Nobel committee awarded the 1992 peace prize to Rigoberta Menchu, a Guatemalan Quiche Indian who has spent the preceding decade denouncing the abuses of her government against the indigenous people.

The international capitalist system created, according to the *dependencistas*' s critique, the growing disparity between the developed nations and the underdeveloped nations and kept the Third World dependent upon the First. One of the most influential of the *dependencistas*, Andre Gunder Frank, argued that the development of capitalism generated, in turn, both economic development and economic underdevelopment. Accordingly, the Third World nations were dependent upon Western finance, technology, and markets while at the same time they were relegated to feeding and fueling Western industrialization through low-growth, low-tech economies that were grounded in agriculture and mineral extraction.

The socialist paradigm presented Third World nations as victims and completely discounted any responsibility on their part for their own history or for seeking cooperative relationships with other Third World nations. Tony Smith, for example, makes it clear that dependency theory overestimated the North and underestimated the South:

> This is not to deny that northern power is real in the South, nor to dispute that its effect may be to reinforce the established order of rank and privilege there, nor to suggest that imperialism is a term altogether lacking in meaning today. But it is to assert that dependency theory has systematically underestimated the real influence of the South over its own affairs, and to point out the irony of nationalists who have forgotten their own national histories. (1986, 27)

The leaders of Southern nations have come to a similar recognition. A report of the South Commission, chaired by Julius Nyerere, the former president of Tanzania, emphasizes the responsibility of Third World nations for their own development and for South-South interdependence and solidarity. "Development," the commission concluded, "is based on self-reliance and is self-directed; without these characteristics there can be no genuine development" (South 1990, 11). With these insights, developing nations are moving beyond dependency theory.

SOVIET DISUNION

Events in eastern Europe and the former Soviet Union have even further weakened the stand of the dependency theorists. With what could only be called the Russian Revolution of 1991, the Soviet empire and the Soviet Communist party were swept away. Political analysts will be kept busy for years to come analyzing the causes of this remarkable change. Even now, however, we can see that the change began two or three years earlier with the reforms of *glasnost* and *perestroika*, the reunification of Germany, the subsequent moves toward independence of some of the Soviet republics, and the failure of the Communist leaders to keep food and durable goods on store shelves. All of this sparked a reactionary coup against Soviet President Mikhail Gorbachev by Communist party hardliners in Moscow.

Having lived almost seventy-five years under Communist dictatorship, the people of Moscow took a stand against the Soviet tanks and for their elected Russian president, Boris Yeltsin. They crushed the coup, toppled the old statues of Communist party leaders in Moscow and the hurriedly rechristened St. Petersburg, and literally shouted out in the Russian parliament for faster democratic reforms and political accountability of their leaders. Revolution and history unfolded on our television sets in one month, August 1991. It was primarily a revolution of the large cities and not the outlying villages, but this was also true of the October Revolution of 1917 that brought the Communist party to power.

Following the August coup and the popular backlash in the cities against the Communist party, Gorbachev futilely attempted to rationalize a continuing place for the party in Soviet politics by appealing to the right for pluralism in a "democratic" society. He was correct when he addressed the Russian parliament and said that socialism was not equivalent to Marxism or Leninism, and that, as an idea, it had support throughout history and even from mainstream religious traditions, such as Christianity.

For political and economic analysts, though, Soviet communism was the principle expression of the Marxist ideology. China remains a hard-line Communist state, but its gray, austere ranks, its undisguised oppression, as evidenced in the Tiananmen Square uprising in 1989, and its Eastern culture make the Chinese brand of socialism certainly difficult to sell to the Third World nations of Latin America and Africa. Other countries, such as Sweden, claim democratic socialism, but this is already a mixture of state control and Western entrepreneurship. The Western industrialized nations, particularly the United States and Great Britain, made it clear that future financial and technical aid to the fifteen republics of the former Soviet Union, if any union does survive, will be determined by how quickly they are able to create a market economy — how quickly, in George Bush's words, they are able to become "more like us" ("MacNeil Lehrer" 23 August 1991).

Variations on the capitalist paradigm of development are likely to persist and may even be given new muscle with the recent turn of events in the former Soviet Union and U.S. insistence on open markets for overseas money. It has become increasingly clear, however, that both the capitalist and the socialist paradigms have been inadequate so far to explain and direct development efforts in the Third World.

KNOWLEDGE AND PORK BELLIES: CONCEPTS VERSUS COMMODITIES

The deficiencies of growth theory and dependency theory become even more glaring when the proposed technology for development is computerized knowledge systems. These theories deal with economic constructs: commodities, production, consumption, and markets. Expert systems are more than hardware parts and market hype. Knowledge engineers profess, at least, to incorporate human knowledge and rules of inference within the expert systems.

Knowledge, in the deepest sense, is not a commodity. In many cases, our commodities make us less human, less free. Knowledge is indispensable to human development and freedom. Clearly, however, knowledge has routinely been regarded as a commodity — as something that can be bought and sold, inventoried and stored, traded and leveraged. It has had economic and political buying power. In the United States, there are laws and patents and protocols to prevent certain knowledge from becoming known outside of the corporate office or outside of the country or outside of the White House.

In reality, knowledge does not exist outside of a human interior. It is a recognition process within that interior. Even though we may structure and stockpile information in libraries and computer data bases, knowledge must still be gathered in by each individual self and uniquely patterned into a new personal consciousness. As important as this new technology may prove to be, expert systems do not have knowledge in the sense that a human person has knowledge. But expert systems and computer technology in general are clearly changing the way we store knowledge, acquire knowledge, and even think about knowledge.

Classical development theories simply do not offer us any meaningful way to talk about computerized knowledge systems in the Third World context. In the decade of the 1980s, these theories came under attack particularly from environmentalists. One response to the inadequacy of these prevailing theories was "green development."

THE GREENING OF DEVELOPMENT THEORY

The United Nations set up, in 1983, the World Commission on Environment and Development, headed by Gro Harlem Brundtland, the prime minister of Norway. The published conclusions of this commission, *Our Common Future*, came to be known as the Brundtland report. In its opening pages, the report acknowledges that economic issues are inseparable from environmental issues. "Poverty," the report states, " is a major cause and effect of global environmental problems" (World Commission 1987, 3). The commission focused on one theme: sustainable development. Sustainable development is defined as development which "meets the needs of the present without compromising the ability of future generations to meet their own needs" (World Commission 1987, 8).

The report admits that developing nations are forced, by necessity, to play the game now by different rules than the developed nations. The distinction is:

> These developing countries must operate in a world in which the resources gap between most developing and industrial nations is widening, in which the industrial world dominates in the rule-making of some key international bodies, and in which the industrial world has already used much of the planet's ecological capital. This inequality is the planet's main "environmental" problem; it is also its main "development" problem. (World Commission 1987, 5–6)

Inequality, according to the Commission, reaches to the very roots of the economic system. It is a system that extracts more from poor nations than it puts in, that straps them with debt, forces them to overuse fragile oils, enacts trade barriers that make it difficult for them to sell their goods at fair prices, and burdens them with inadequate aid that reflects the priorities of the rich nations rather than the poor (World Commission 1987, 6). In spite of all this, the Brundtland report was optimistic that poverty could be eliminated and that the earth could sustain increasing industrialization. The message of *Our Common Future* was that "technology and social organization can be both managed and improved to make for a new era of economic growth" (8).

The Brundtland report itself came under fire from environmental groups. Some of these criticisms are taken up by Thijs De La Court in *Beyond Brundtland: Green Development in the 1990s*. Environmentalists were disturbed by the Brundtland report's optimism that continued large-scale industrial growth was not only possible but necessary. De La Court quotes Gandhi who, when asked by a British colonial whether newly independent India would achieve Britain's standard of living, said: "It took Britain half the resources of the planet to achieve this prosperity; how many planets will a country like India require?" (1990, 15).

Six principles are proposed by De La Court to describe the type of development that environmentalists can support (1990, 136). They are the basis of what has come to be called "green development":

1. Development must come from within not from without.
2. Development must be compatible with and restore diversity of species and depend on sustainable resources.
3. Development must provide the basic human necessities for all people and promote equity.
4. Development must promote self-reliance, empowerment, and grassroots participation.
5. Development must be nonviolent, in both the personal and institutional sphere.
6. Development must allow for mistakes without endangering the ecosystem and resources.

Green development has, in its turn, also come under close scrutiny. William Adams gives a thorough history and critique of the movement. According to Adams, the movement is defined by the concerns of environmentalists for global issues — such as acid rain, the ozone hole, the greenhouse effect, rainforest depletion, loss of wildlife habitat — and attempts "to move beyond environmental protection and transform conservation thinking by appropriating ideas and concepts from the field of development" (1990, 1). Its key rhetorical phrase, taken from the Brundtland report, is "sustainable development." Adams regards this as a better slogan than a basis for theory because it "suggests radical reform without actually either specifying what needs to change or requiring specific action" (1990, 4). He marshals Timothy O'Riordan's argument that the concept of sustainable development combines the rationalism of the technocentrist approach with the romanticism of the ecocentrist approach (1990, 12).

Apart from its focus on environment and ecology, the real importance of the green movement in the unfolding of development theory lies in its concern for the empowerment and self-determination of the poor. Green development looks beyond centralized planning to grassroots participation or "development from below." Both the capitalist paradigm and the socialist paradigm for development were predicated in elitism. In the capitalist paradigm, the elites were the economists, the technocrats, the scientists, the experts. In the socialist paradigm, they were the socialist party leaders. Both approaches encouraged authoritarianism, hierarchies, and central planning.

While green development, according to Adams, is often, though not always, based on an "essentially libertarian position" and disparages hierarchical structures, it does recognize limits set by technology and social organization and "throws attention back on the ethical questions" of development (1990, 55, 59, 199). "Development," Adams writes,

> ought to be what human communities do to themselves. In practice, however, it is what is done to them by states and their bankers and "expert" agents, in the name of modernity, national integration, economic growth or a thousand other slogans. Fundamentally, it is this reality of development, imposed, centralizing and often unwelcome, that the "greening" of development challenges. (1990, 199)

Empowerment has become almost a catchword in this decade of the 1990s. Television talk-show guests speak facilely about empowering abused women or empowering consumers. Green development, for all its ambiguity, is about empowering the poor. Adams makes this very clear when he concludes his critique with the statement that the "greening" of development "involves not just a pursuit of ecological guidelines and new planning structures, but an attempt to redirect change to maintain or enhance the power of the poor to survive without hindrance and to direct their own lives" (1990, 202).

Pierre Pradervand is both an articulate spokesman for grassroots development and a man who challenges Westerners to recognize the knowledge, wisdom, and experience of the world's poor. One of the main failures of development theory and development practice in the last fifty years, for Pradervand, is that it has excluded the people from participation. "'Development,'" Pradervand claims, "has been something that has been done for people, to people, sometimes despite them and even against their will, but rarely with them" (1989, 22).

Pradervand spent almost five months traveling to 111 villages in five countries of Africa: Senegal, Mali, Burkina Faso, Zimbabwe, and Kenya (1989, xiv). During those journeys he talked to approximately 1,300 farmers about their village projects and cooperatives, and documented their creativity and courage, their hopefulness and wisdom in the face of what, to us in the West, seem insurmountable problems and crises. From his personal experiences with Africa's traditional farmers, Pradervand derived his own definition of "meaningful" development: a "process of empowerment" that allows each person to "express his maximum potential" (1989, xvii).

This new approach to development has richer possibilities than growth theory or dependency theory when it comes to a discussion of expert systems in the Third World. It places an emphasis on individual and community knowledge and how that knowledge can be used to develop the poor rather than to control them. Pradervand was writing about self-help peasant groups in Africa, not knowledge engineers, but the approach that he proposes could also be useful for those seeking to implement expert systems in a Third World community. Development begins, Pradervand urges, with an appreciation of what people are (their identity), a respect and sometimes a rediscovery of what they know (their traditional knowledge and techniques), and an effort to learn what they wish to achieve (1989, 22).

These concepts of empowerment and participation of the poor, are not, in themselves, a theory of development. They are simply working guidelines, and they only get us partly to where we are heading. What we require is a development ethic.

ETHICS AND DEVELOPMENT

Knowledge shapes our intellectual, our aesthetic, and our moral being — the human interior from which choice, love, and self-reflection emanate. It distinguishes the great civilizations from the unorganized masses struggling for life and human dignity. It makes moral choice possible. Knowledge, for example, must always precede moral decision and in some cases it compels moral decision.

There are clearly ethical issues involved in withholding, in sharing, and in using knowledge. With 40,000 children dying every day in the Third World, is it ethical to sell the medical knowledge or technology to prevent some of these deaths? Is it ethical to profit from the Third World's need for knowledge and expertise? Is it ethical to sell knowledge systems to Third World nations that will be used internally to control populations or to secure the position of an elite group? Is it ethical to promote expert systems if they necessarily create social divisiveness by valuing one kind of knowledge over another?

The technology of expert systems necessarily defines who is judged to know and who is not, what knowledge is valued in the society and what is not. And since the National Research Council also sees expert systems as playing an especially important role in national planning for developing countries, these systems may also define which changes in the environment and the society should be priorities and which should not (1988, 20).

NOBODY SEES THE WIZARD — NO WAY, NO HOW

The Random House Dictionary of the English Language defines an expert as "a person who has special skill or knowledge in some particular field; specialist; authority" (1987, 681). Authority is subsequently defined as "the power to determine, adjudicate, or otherwise settle issues or disputes; jurisdiction; the right to control, command, or determine" (1987, 139). Aside from the fact that skill, knowledge, and authority have generally been assumed, up until now, to be qualities only ascribed to a person, from where do expert systems get their authority, their power? Short of sending in a robotic Toto, how do we lift the curtain and see the wizard at the controls?

An expert system is derived from many sources: the expert in the specific area that comprises the expert system, such as a physician or a chemical engineer; the knowledge engineer who extracts the information from the expert; the programmer who codes the expert program according to the system designer's specifications. Presumably, these could be one person but it is more likely to be three or more — often many more. When computer software comes out of a research laboratory, its "author" is sometimes known. Terry Winograd, for example, was known in AI circles as the one who created SHRDLU, a computer program that simulated on a video screen a robotic arm moving toy blocks in response to commands issued from a keyboard. But, for the most part, computer software is anonymous. It is also frequently modified by others who were not part of the original design or programming effort. Who takes responsibility, therefore, for the network of inferences that are built into an expert system and upon which decisions are made?

FOR EXPERTS ONLY

Computerized knowledge systems have the potential to affect political, cultural, and social decisions and relationships. They can alter the storage of knowledge, reasoning about knowledge, the uses of knowledge, and the very concept of what is knowledge — and who qualifies as an "expert." Originally, expert systems were not designed to replace experts. They were designed for experts. MYCIN, one of the prototype expert systems that, like DENDRAL, grew out of the experiments in heuristic programming at Stanford University, was principally designed by Edward Shortliffe to be used by doctors prescribing antibiotics and to offer them suggestions and reinforcement for their therapeutic decisions. At the time work was begun on the program in 1972, there was evidence that doctors were overprescribing antibiotics and were sometimes failing to even identify the organism that was the source of the infection (Buchanan & Shortliffe, 16).

Today you still find in the literature descriptions of expert systems for the expert — for example, an expert system to assist paleontologists in identifying microfossils found during exploratory oil drilling (Swaby 1992, 36). An investigation of expert systems in the workplace — at DuPont, Northrop, IBM, Texas Instruments, Fujitsu, and other corporations — found that expert systems were never implemented as stand-alones; they were assistants to the decision makers and professionals, not replacements (Feigenbaum, McCorduck, & Nii 1988, 8).

If only for that reason, though, expert systems are likely to enhance the positions of those already in power within the Third World nations, whether that power resides in government, in academic institutions, in medical institutions, or other centers of knowledge. It is always the educated and the affluent that have access to the latest technology — be that printed books, image-enhanced telephones, heart transplants, in vitro fertilization, or personal computers. Yet, even a peasant farmer or a local fisherman has knowledge about his environment and approaches to ensure survival. Poverty or lack of formal education does not equate to lack of knowledge or lack of technology.

POWER IN THE MODEM AGE

Not everyone, of course, agrees that "knowledge is power." Plato, for example, desired to create the philosopher-king because he realized that those who sought wisdom were generally not those who wielded power. However, computer technology has given us new ways of looking at knowledge and rationality. Walter Ong once told me that, when an optical scanner was used to input a printed manuscript of his into a computer file, the system changed the phrase "modern age" to "modem age." A second attempt yielded the same result. I suggested that he should have listened to the computer the first time because it knew what it was doing. In this modem (postmodern) age, there is a sense in which knowledge is power and in which knowledge systems can be used to empower or to oppress. Evaluation of expert system technology, therefore, requires a theory of development based on ethical, not economic, tools of analysis.

JOHN RAWLS ON JUSTICE

Those tools will not be easy to find. In the aftermath of Third World disasters, such as the Bhopal, India, chemical leaks, there has been renewed interest in the issue of ethics within technological development. Some writers have turned to John Rawls's "two principles" as a "duty ethic" to guide multinational corporations operating within Third World nations and the governments of those nations. The most recent formulation of the two principles is given by Rawls as follows:

1. Each person has an equal right to a fully adequate scheme of equal basic liberties which is compatible with a similar scheme of liberties for all.
2. Social and economic inequalities are to satisfy two conditions. First, they must be attached to offices and positions open to all under conditions of fair equality of opportunity; and second, they must be to the greatest benefit of the least advantaged members of society. (1987, 5)

The difficulty here is that these principles assume free, autonomous persons, acting rationally. All of these terms — liberty, freedom, person, rationality — become ambiguous in the Third World context.

The terms themselves are interrelated in Rawls's definition. He defines autonomy as "acting from principles that we would consent to as free and equal rational beings, and that we are to understand in this way" (1971,

516). One's moral convictions about justice, according to Rawls, cannot be the result of "coercive indoctrination" from those in authority. In actual fact, the moral convictions of the vast majority of us are instilled by the authority of institutions, teachers, parents, and the media; it becomes very difficult to determine what is indoctrination and how our genes, our education, and our culture have limited or liberated us. On a global scale, the military, political, and economic institutions of the First World exert tremendous pressure upon the people and institutions of the Third World, determining, in some cases, their government leaders, their national borders, their exports and imports, their development policies, their education, and their personal and social decisions.

Even if the "least advantaged" of the world community assented to the inequality of their conditions in order to maximize the benefits of the relationship, the assent would be constrained by all these pressures. In fact, the ambiguity of the principles leaves them open to an interpretation that justifies an elite establishment as long as the "least advantaged" continue to benefit.

Rawls is addressing the question of justice within a Western, democratic, modern, individualistic society. He clearly acknowledges that his principles of justice are in the spirit of Immanuel Kant, who argued that moral principles come from the rational choice of free and equal human beings. In his later writings, Rawls admits that his principles may not be universal and, therefore, may not be appropriate for other societies:

> An immediate consequence of taking our inquiry as focused on the apparent conflict between freedom and equality in a democratic society is that we are not trying to find a conception of justice suitable for all societies regardless of their particular social or historical circumstances. We want to settle a fundamental disagreement over the just form of basic institutions within a democratic society under modern conditions. . . . How far the conclusions we reach are of interest in a wider context is a separate question. (1980, 518)

Even his toughest critics, such as Robert Nozick, recognize the importance of Rawls's contribution to ethics. What development theory requires, however, is something that goes beyond Rawls: a global, intersocietal conception of justice.

THE HUMAN ASCENT

There are, however, other emerging voices that might provide at least an approach to global development, if not a comprehensive theory. One of these is Denis Goulet. In the 1970s, he made an early attempt to look at the concept of development and at ethical strategies for development, and has followed through with this study in the intervening years. Development ethics is a study, Goulet says, which requires us to "systematically define the symbolic and institutional requirements of the good life, of the just society, excluding domination and exploitation in a world of convulsive technological change" (1974, 2).

As such, development ethics require self-reflection and self-criticism by the First and Third World alike. Moreover, Goulet at least alludes to the limitations of both knowledge and cross-cultural transfer of knowledge in the development process. Like Thomas Kuhn, Goulet proposes that new knowledge is rarely the "creation" of individuals working in isolation and rarely free from ethnocentrism. Gestating in communities engaged in dialogue, knowledge is "derived from limited, personal, cognitive experience in a given cultural mode" (1974, 18).

Attempting not to ignore the economic component of development but insisting that links between economics and moral philosophy be restored, Goulet contends that economics, caught in a general trend toward specialization of knowledge, "has achieved great virtuosity in handling means, but it is no longer competent to evaluate ends" (1974, 10). This is a theme also taken up recently by Amitai Etzioni. Etzioni's position is that individuals are able to act rationally and advance their self or "I," but their ability to do so is affected by how well they are anchored and sustained by the moral values of a community that is perceived as "we" and not as a restraining "they" (1988, ix–x). All this is in contrast to the paradigm of neoclassical economics, which is utilitarian, rationalistic, and individualistic; the individual seeks to maximize his or her utility and pleasure by rationally choosing the best means to his or her ends.

Goulet was greatly influenced by Louis-Joseph Lebret, who insisted that it was useless to theorize about development until one first experienced the shock of underdevelopment. Goulet comes back to this same point again and again in his own reflections. The shock of underdevelopment is "traumatic, existential contact with mass misery" (1974, 30). This is just the opposite of "scientific objectivity." It is seeing underdevelopment from the side of those we have conveniently labeled underdeveloped. For the poor, Goulet writes, "underdevelopment is a sense of personal and societal impotence in the face of disease and death, of confusion and ignorance as one gropes to understand change, of servility toward men whose decisions govern the course of events, of hopelessness before hunger and natural catastrophe" (1971, 23). The poor are vulnerable before our power, our knowledge, our technology, our paternalism, and even our good intentions.

Goulet has similarly adopted Lebret's conception of development as a "human ascent" of ongoing transitions from the less human to the more human (1974, 35). Lebret did not deny that rational planning, business investment, reformed institutions, and political mobilization of the poor can contribute to development. He simply argued that they are not sufficient to achieve "authentic development" unless there is a "cultural revolution" that changes human values. It is specious for those in the Western industrial nations, for example, to demand more oil for their cars, cheaper labor for their relocated factories in Mexico or Thailand, cheaper goods for their stores, more technology in their hospitals and their homes, if these escalating and created needs keep millions of other people in poverty and suffering. A development ethic demands that we globally prioritize needs. It also entails the cultivation of a discipline or spiritual capacity so that, as we move beyond basic needs or as our needs change, we can "have more" without "becoming less" human (1974, 39).

ENCYCLICALS ON SOCIAL JUSTICE

These ideas are echoed in some of the Catholic encyclicals which address the issues of human work and social injustice: *Rerum Novarum* (1891), *Mater et Magistra* (1961), *Populorum Progressio* (1967), *Octogesima Adveniens* (1971), *Redemptor Hominis* (1979), *Laborem Exercens* (1981), *Sollicitudo Rei Socialis* (1987), and *Centesimus Annus* (1991). Their importance for our purposes here is not as "religious" documents. The Catholic church is a global organization and the majority of its members live in Third World nations. Within its organization, it has had to reflect on human rights and duties; to see the poor; to accept responsibility for its own silence or obstruction of social justice; to acknowledge in its history its own oppression of "conquered" nations and cultures; to look upon its own surviving inequities of wealth, power, color, and gender; and to weigh what appears to some as its own unrelenting and unnecessary suppression of challengers within and without its ranks. Light struggles with darkness — in individuals and in institutions.

The light certainly comes through, however, in *Populorum Progressio* (On the Development of Peoples) in which Paul VI wrote candidly about the global "scandal of glaring inequalities" in both possessions and power that he had personally observed during two journeys to Latin America and Africa before he became pope (1967, 7). The encyclical warns that "development cannot be limited to mere economic growth" and that "avarice is the most evident form of moral underdevelopment" (1967, 10, 13).

John Paul II further develops the concept of a "duty of solidarity" in *Sollicitudo Rei Socialis* (On Social Concern). Describing development as "a duty of all towards all," he, in effect, denies that there are different "worlds" and says that, when development in one of the so-called worlds is at the expense of the others, "development becomes exaggerated and misdirected" (1987, 58). In *Centesimus Annus* (The Hundredth Year), John Paul II speaks also of a new kind of ownership and wealth that separates the rich and poor nations: the ownership of knowledge and technology (1991, 13). Significantly, the concept of property, which originally included land and its produce, now embraces what is intangible and cannot be neatly counted, measured, and transferred to a new buyer.

Simplicity, discipline, sacrifice — urgings to these have always met with resistance, whether the speaker was Jesus of Nazareth, Mao Tse-tung, Mahatma Gandhi, or Julius Nyerere. None of these were naive men. In fact, their ability to mobilize and lead hundreds, thousands, even millions of followers is an indication that they were politically astute men. For the most part, those writing about development theory have assumed that poverty and injustice were something that could be "fixed" with more money, more technology, more reforms of political and social institutions. In the name of scientific objectivity, there was an effort to keep ethical values out of the discussion even though all of these — money, technology, and institutions — embody literally and symbolically the values of a society and culture.

Consequently, those who look upon expert systems as just another tool of Western culture are mistaken. As a repository of knowledge, rules of inference, and simulated expertise, expert systems are quite different from the human tools with which we have so far shaped our environments and ourselves. They are potentially a powerful and deceptive tool. We have often in the West exported our tools, gone to war, entered into alliances, and manipulated international trade and finance without applying ethical values to our artifacts or our actions. But to the degree that we have done that, we ourselves have been humanly underdeveloped. A development ethic is required not simply to protect the weak but to redeem the powerful as well.

Chapter 4

Wanted: Experts for Third World Development — Thirty-Five Megabytes of Experience Required

An expert is a damned fool a long way from home.

— Carl Sandburg

They are wrong about facts, they are wrong about theories, they are wrong about dates, they are wrong about geography, they are wrong about the future, they are wrong about the past, and at best they are misleading about the present, not to mention next week.

— Christopher Cerf, *The Experts Speak*

You cannot create experience. You must undergo it.

— Albert Camus

If the Third World is popularly viewed as in an incessant state of crisis, expertise is most often held up as the solution to those crises. Western governments and corporations are eager to market their machines, processes, and consultants in Africa, Latin America, and Asia. Third World governments have been just as eager to acquire Western technology and recruit Western experts.

Expert systems have added a new dimension to this concept of technology transfer for accelerated development of the Third World. The technology that is being exported is no longer simply a product of Western expertise. With expert systems, Western expertise itself has become the technology.

It is true that Western experts and consultants have, for many years now, peddled their know-how in the Third World marketplace. Very often the technologies they were marketing were intangible and had to do with ways of interpreting and organizing information, with decisions, diagnoses, and management.

Certainly, technology is embodied in persons as well as in products and processes. In a sense, all technology is embodied in persons because it is first conceived in the human mind and its intended purpose is held there. It is not unusual to come across an item in an antique store, for example, and puzzle over its former use. When its use is no longer apparent or discernible or the context in which it must be used is no longer existent, it has ceased, in a real sense, to be technology.

There have been a few multipurpose tools, such as the Swiss army knife or amphibious vehicles, but, for the most part, tools have been designed for specific purposes, and the designers had these purposes in mind. The computer is a unique tool because its purpose is constantly being reinvented by its users. Its power consists in the fact that it is a symbol machine, and its symbols and their interpretations can be altered.

In earlier forms of technology, whether that technology was of a tangible nature, such as machinery, or an intangible nature, such as formulas or procedures, it was a human person that was applying human rationality to the solution of problems. Expert systems, however, have changed the game. They are purportedly the distillation of a specific area of knowledge in which the rules of rationality and inference are encoded in computer software. In expert systems, the human person can, theoretically, externalize not just his or her conception for a technology to solve an existing problem but the reasoning ability to solve problems that have not even been posed yet.

KNOWING MORE AND MORE ABOUT LESS AND LESS

During a commencement speech in the early 1940s, Nicholas Butler told a graduating class at Columbia University that they were now certified experts in their field. It was a left-handed compliment because he then explained that "an expert is someone who knows more and more about less and less." Perhaps every culture takes aim at its experts. The Sufis, a mystical sect of Islam, have a tale about a man who suddenly revives as his friends and neighbors are about to nail shut his coffin. The man rises up in protest, but since the local experts — the doctors and the priests — have certified his death, he is shoved back into the box and duly buried. For the moment, I want to focus on the concept of expertise itself because existing Western attitudes toward expertise have emerged from our unique technological environment.

Admittedly, we live in a culture that uses the term *revolution* rather offhandedly. It can mean anything from the bloody rise and the wearied fall of Soviet communism to a lose-fat-while-you-sleep diet pill. Some speak of a "revolution" in Third World development (National Research Council 1988, 19). The green revolution, which introduced new seed varieties and agricultural techniques to the Third World, has supposedly given way to the "gray revolution." The term does not refer to the political mobilization of the over-55 crowd. Rather, it refers to the gray matter of the human brain and the contention that human reasoning can be incorporated into computer software and used to approach development problems.

There is clearly the suggestion here that development can largely be reduced to an explicit definition of problems and the application of the appropriate expertise to eliminate the problems. There is also the suggestion that, in the course of development, knowledge or solutions are imparted from those who know to those who are judged not to know. Both of these assumptions need to be examined because they underlie so much of the current thinking about expertise and Third World development. They have also begun to steer the early discussions about expert systems in the Third World.

Some of this examination has already begun with writers such as Denis Goulet and Paulo Freire. Neither of these individuals has actually addressed the question of expert systems. For the most part, their interests, along with the other writers that I cite in this chapter, lie in ethics, education, and technology. In many cases, their reflections and writings came from before expert systems were even a gleam in the eye of an AI researcher. Nevertheless, some of their insights have direct relevance to the issue of expert systems in Third World development and will help us to see that issue from more than just a technical perspective.

When we talk about transferring technology to the Third World, we need to remember, of course, that all societies have their indigenous technologies, whether these are the boomerangs of the Australian aborigines, the dugout canoes of the early Polynesian navigators, the dung-burning stoves of rural Africa, or the microcomputers and space shuttles of the United States. Each of these technologies represents a pragmatic solution within a particular social context. Generally, they are more than this. They can represent aesthetic, ethical, political, and economic choices as well.

In Denis Goulet's analysis, human societies have an "existence rationality," which comprises "the strategies employed by all societies to process information and make practical choices designed to assure survival and satisfy their needs for esteem and freedom" (1971, viii). Consequently, the technology that a society develops is largely determined by its particular existence rationality.

INSTRUMENTS OF COMPLEXITY

Until quite recently in human history, technology has been simply one among many social strategies for information processing that have also included authority structures, kinship relationships, apprenticeships, proverbs, rituals, myths, and religious revelations and prescriptions. However, in the Western nations, technology has rapidly become the pervasive mode of information processing and problem solving.

Any of us could come up with familiar examples to illustrate this observation. Chemical analyzers do blood profiles on patients admitted to the hospital and indicate to the physician possible areas of disease. Radio transmitters and computers process signals from distant galaxies and provide astrophysicists with information on the origins and nature of the universe. Television has become the main source of news and entertainment for most Western families and subtly shapes their image of themselves, the "world," their enemies, their wished-for lovers, their fears and desires. Computers model the economy, the next nuclear war, facial reconstruction for the patient about to undergo plastic surgery, the outcome of national elections — at times, it seems, just about everything that calls for a major decision in our high-tech lifestyle.

Technology, therefore, is widely used in Western society to gather our information, disseminate it, define our problems, and determine our solutions. More often than not, we in the Western nations defer to the perceived experts. When the area of investigation is especially complex or time-consuming, such as the mathematical calculations to guide Voyager II on its decade-long journey through the planetary system and out into the sea of stars, then the experts themselves defer to computers. Heinz Pagels refers to the computer as the "primary research instrument of the sciences of complexity," since the computer is able to manage vast amounts of data and simulate reality (1988, 13). "We may begin to see reality differently," Pagels argues, "simply because the computer produces knowledge differently from the traditional analytic instruments" (1988, 13). Clearly, we have already begun to see and think of ourselves differently. This is true not only in new research programs, such as cognitive science which models the human mind on the computer, but in our everyday conversation: "Can I access your files on cold fusion?" "He was programmed for failure." "We are suffering from information overload."

PROBLEM SOLVING

Moreover, Western technology is reductionist in its approach to rationality. According to the modern technological mind-set, to be rational is to experience reality as a problem that can be analyzed, broken into parts, fixed, reassembled, manipulated, and measured for effectiveness and productivity.

Technology has become not simply the depository of Western society's practical knowledge but has, in Denis Goulet's expression, replaced nature "as the context of societal perceptions and decisions" ([1977] 1989, 31). This is a theme that runs through the work of Jacques Ellul as well. The forces that Western society must now confront are not primarily wind and rain and fire, but the forces created by technology itself — electricity, machinery, infrastructures, modes of operation, computerized decisions, life-support systems, "smart" weapons, the virtual reality of television and computer games, electronic sound systems, and so on.

"The global diffusion of modern technology," Goulet goes on to say, "tends to standardize the 'existence rationality' of all societies around specifically Western notions of efficiency, rationality, and problem-solving" ([1977] 1989, 16). On the evening news many Americans look at a stumbling Russian economy, a starving Somalian child, or an Islamic demonstration in Iran against the "Satan" of the United States, and shrug, "Why can't they be more like us?" The implication is that they want the same things that we want, but they are not as good or smart as we are in getting them.

In the non-Western society, on the other hand, notions about efficiency, rationality, and problem solving may mean something entirely different. For the Muslim, the efficient way to work might be one which allows him to recite the prayers of the Koran seven times a day and reduce physical exertion during the periods of prescribed fasting, such as Ramadan. To the Hindu, rationality might encompass a profound experience of the "reality" of the great myths about Krishna and Vishnu. To the San, drought might not be a "problem" to be solved but an occasion for reflection on the necessity of ongoing harmony with nature or of human endurance.

CRUCIBLE OF CHANGE

This is not to imply that all Western technology and all Western contact is damaging for Third World nations. A living culture cannot be preserved intact. All cultures are constantly changing as the immediate environment is altered and cultural exchange is accelerated by electronic communication. The consciousness of Third World nations has been awakened. They have seen, through television and advertisements, the differences between their own lifestyles and that of the Western world. This new consciousness creates new aspirations, new consumer wants, and new expectations in communities of the poor around the world.

Goulet has, nevertheless, called development a "cruel choice" because its benefits are purchased at a high price and the poor are not guaranteed happier or freer lives after they have paid the price (1971, 326). The benefits of development can be alienating rather than liberating.

On the other hand, Western nations have only focused on the problems of Third World nations. They have generally failed to see the riches in these nations and the human poverty in their own. Erich Fromm has rightly observed that if there is an alienation in misery, there is also an alienation in affluence. That alienation is evidenced in the United States, for example, in a promoted militarism to the detriment of social programs for health, education, and urban renewal; in racism; in escalating rates of drug use and drug addiction; in consumerism; in domestic violence; in rising rates of divorce, teenage alcoholism, and teenage suicides; in child abuse; in the growing gap between rich and poor; in voter apathy; in children killing children with handguns; in sexual irresponsibility with AIDS as the gamble and abortion as the solution; and in other indices of problems within our own society that our technology has not solved — and, in some cases, appears to have created or worsened.

This recognition often comes as a shock. Goulet challenges Western smugness with the observation that "it is discomforting for a sophisticated technical expert from a rich country to learn that men who live on the margin of subsistence and daily flirt with death and insecurity are sometimes capable of greater happiness, wisdom, and human communion than he is, notwithstanding his knowledge, wealth, and technical superiority" (1971, 27–28). Until we come to this recognition, our overtures to provide technical expertise — including expert systems — to Third World nations remain part of the problem of underdevelopment.

Development and underdevelopment are, at base, power issues. The experience of the Third World is, in part, that of powerlessness and vulnerability. Their governments and strategists perceive that they must acquire Western technology in order to overcome that powerlessness and vulnerability. Unfortunately, the symbols of Western power which many of these nations first turn to are our guns and fighter planes and missiles. Disproportionately large amounts of Third World income is spent on military hardware.

The aspirations of the Third World community for power and equality in the international arena of money and politics are largely futile, however, since the nations of the West have never intended power to be transferred with their technology. That is one of the main reasons for the stringent legal restrictions on what technology can be transferred and to whom.

It is also why the United States has been quick to act in preventing nuclear proliferation, particularly in the Middle East. Apparently, with the political justification that the United States and its Western allies alone have the moral superiority to deploy these arsenals responsibly and to protect the "new world order," the U.S. government has been vigilant in ferreting out nuclear research plants in Iraq, Iran, Libya, and other nations of the Third World. When the Bush administration learned that both Iraq and Iran might be closer to "the bomb" than their CIA network had suspected, a government source is reported to have said, in defense, that previously no one thought the "ragheads" were able to build a nuclear bomb ("NBC Nightly News" 11 July 1991). "Ragheads" was the derogatory term used for the Arabs. Its demeaning tone was on a par with President Bush's reported comment, during Saddam Hussein's occupation of Kuwait, that the United States was "going to kick ass." Such off-the-cuff rhetoric, reported nevertheless on national television and in the leading U.S. newspapers, is more revealing of Western attitudes toward the Third World than a library of official documents and press releases.

APPROPRIATE TECHNOLOGY

More than Western attitudes are important here. Western technology is the bearer of Western values. Some, of course, regard all technology as neutral, something value-free in itself that can be used for good, bad, or amoral purposes. I would argue with others that technology emerges from a cultural setting and reflects the life-values of that setting. In his book, *The Culture of Technology*, Arnold Pacey writes that we would do well to see technology "not only as comprising machines, techniques and crisply precise knowledge, but also as involving characteristic patterns of organization and imprecise values" (1983, 4).

Life-support systems are a timely case in point. Developed and widely used in the West, they are the technology to keep the heart and lungs functioning when brain activity can no longer do this effectively and to nourish the patient with intravenous feeding. In the case of severe and irreversible brain damage, they are a means of conserving the ecosystem of the body. It has become common to read in our newspapers about a parent, a husband, or a wife petitioning the courts to have one of their family members either taken off life-support systems or kept on life-support systems. These family crises have been dramatized in television programs and trigger more and more discussion about "living wills" and "right to die" laws. Such discussions question the assumptions of this technology which are, in large part, that physical life must be sustained as long as possible because death is the ultimate tragedy for the individual and the ultimate failure for the physician. These are the values of a materialistic culture and are at odds, for example, with the professed values of genuinely Christian, Islamic, Buddhist, and Hindu cultures.

The beliefs and values that precede a technology, that surround its development and direct its use are often unexamined and tangled — difficult to get at and articulate. But they are there. Some of these are beliefs about technology itself. It makes a tremendous difference whether technology is conceived of as an endless pursuit and exercise of technical virtuosity, efficiency, complexity, power, and market hype, or whether it is perceived as the compassionate use of skill and innovation to remove the remediable causes of suffering, fear, ignorance, loneliness, and servility, and to promote a basis for temperate living.

If the latter sounds more like a description of missionary activity than technology, it is because the prevalent Western understanding of life on earth is at odds, for example, with the enduring religious and humanistic visions that stress these values. To even talk about goodness, truth, and temperance is often seen as an academic embarrassment. Yet, some, such as E. F. Schumacher, have sought inspiration from such values. The rationale for intermediate and appropriate technology, which Schumacher related in *Small Is Beautiful*, came from the Buddhist philosophy of the "middle way" and "right livelihood," as well as from the Christian tradition that the genuinely good life must be realized through prudence, justice, fortitude, and temperance (1973, 280).

Schumacher speaks of "Buddhist economics" but says that the teachings of Christianity, Judaism, Islam, or any of the great Eastern traditions could have been used as well. Each of these is concerned with values and meaning. What is important is that development, including technological and economic development, must proceed from a vision of the human person and the purpose of life and not simply from a theory of production and consumption.

These same humanistic traditions, of course, cease to be integrative when they are approached as simply a collection of dogmas or sayings that can be taken out of context and used to defend a position. Their judgments become fragmentary — like those of science. Many are fond of quoting from Genesis, for example, that "man" was given "dominion" over all the earth and conclude, therefore, that the Judeo-Christian ethic justifies and even promotes rampant technological development and environmental destruction — as well as sexual discrimination.

Americans, especially, have had a longstanding love affair with technology. We have been delighted and fascinated by each new technological toy — from wristband televisions to laptop computers. Even those technological marvels that most of us will only read about, such as atomic accelerators and space telescopes, have seized our imaginations and won our support. Arnold Pacey cites a remark of J. Robert Oppenheimer, who led the project to develop the first atomic bomb, as an example of Western society's "technological exuberance." Oppenheimer found himself in opposition to Edward Teller, who conceived of the hydrogen bomb and promoted its development. Although Oppenheimer realized that the hydrogen bomb, which derived its explosive energy from thermonuclear fusion rather than nuclear fission, would far exceed the destructiveness of the atomic bomb, he came to see Teller's concept as "technically so sweet that you could not argue" (Pacey 1983, 81).

THE OMNISCIENT EXPERT

Behind all this technological development and adulation is the presence of the expert. Beliefs about expertise impact all our technological decisions — not simply the design of our expert systems. For Pacey, one of the dangers of this "cult of expertise" is that the expert has the upper hand in Western technological societies and also wants to exert that same control in the name of Third World development. The expert is in control, according to Pacey, because he or she "can always present knowledge selectively and manipulate people by pre-empting decisions," frequently out of a sense of urgency that "arises from a view of life that has been narrowed by professional training" (1983, 153). Certainly at the village level of development work, the disparity in formal education between the Western experts and the villagers will "tempt the former into believing that existing local technology is of little worth, and that their knowledge as experts is a better basis for planning for the future" (1983, 150).

Denis Goulet has written in *The Uncertain Promise* about the "matrix of technology" and the value conflicts that arise from extracting technology from this matrix and applying it to development in poor countries. One of the ways that technology effects development is in altering modes of decision. Certainly, in expert systems, we propose to alter the mode of decision by placing the rules by which the decisions are made within a computer program. If you believe that relating to nature and human events is a matter of problem solving and that sound decisions can only be reached by experts or by simulated experts on a microchip, then you will end up with a select group of people who will define the problems and reach the solutions for others — either through their own reasoning process or that of a computer program.

CRITICAL CONSCIOUSNESS

It was precisely in response to such a condition that Paulo Freire, in the late 1960s, offered the alternative concept of *conscientizacao* or critical consciousness. He does not abandon the Western approach of problem solving but he does reject the idea that experts must be the ones to solve the problems for others.

By his own admission, Freire has gathered insights from diverse philosophical positions in an effort to respond to the real-life poverty and injustice of Latin America. His analysis certainly echoes with strains from Karl Marx and Martin Buber. Like Marx, he views history as a struggle in which the oppressor is locked in a dehumanizing encounter with the oppressed. Like Buber, he sees the source of dehumanization in the fact that the oppressor regards those he seeks to dominate as things rather than persons. The oppressor possesses the oppressed in his consciousness as an object at his disposal not as a person with whom to enter into dialogue.

Freire's intent in *Pedagogy of the Oppressed* was to define the nature of a liberating education, liberating for the oppressor as well as the oppressed. Obviously, not every twist and turn is clearly mapped out in Freire's critique. The nagging problem has always been: What happens after consciousness has been raised? What happens after the revolution? It is the problem of humanization, of the individual becoming human. There is a mystery and tragedy to human life that no amount of social, political, and economic development can penetrate or avert. Freire is aware of the dangers in the process because he acknowledges that as oppression gives way to liberation, the vision of the "new man" is generally individualistic. The liberated individual aspires to all the apparent rights, freedom, privileges, indulgence, and power that were observed in the oppressor. Too often the tyrant is the slave unshackled.

The practical thrust of Freire's work, nonetheless, was the education of illiterates in Latin America using techniques that would help them come to see the injustice of their situation and the fact that their situation was not inevitable. Because of their lack of education and their position of powerlessness in a society dominated politically, socially, and economically by a group of elites, these individuals belonged to what Freire termed a "culture of silence" ([1970] 1988, 13). They were unaware of their own situation and, therefore, unprepared to change it. The purpose of developing their critical consciousness was to make them aware and to make them see their ability to transform the reality in which they lived out their lives.

What is particularly relevant for our purposes here is Freire's argument that education is not simply a matter of transferring knowledge from those who possess it to those who do not possess it. Knowledge, for Freire, is a matter of relationships between knowing beings and not just between a knowing subject and a known object. "Without a relation of communication," Freire writes, "between subjects that know, with reference to a knowable object, the act of knowing would disappear" ([1973] 1987, 136). The end of knowledge is not the object known but the communication between knowing subjects.

It is not possible, in Freire's view, to transfer knowledge or expertise from one person to another. "Liberating education," Freire writes, "consists in acts of cognition, not transferrals of information" ([1970] 1988, 67). The act of knowing requires dialogue about a problem that leads the person seeking knowledge to critically see the reality in which he or she exists and to make his or her own choices and decisions based on that perception.

For Freire, there can be no communication, no education, no act of knowing, if the relationship at the outset is between an expert and a peasant, one who knows and one who is taken to be ignorant. This is true even if the expert "understands" the peasant's language and the context of his beliefs. Freire uses the example of an agronomist and a peasant farmer in northeast Brazil. The peasant farmer tells the agronomist that he cures infected wounds in his animals by praying over their tracks in the mud. The agronomist understands the farmer's words and the context in which those words are spoken. Still, since the agronomist does not share the farmer's belief, he invalidates "all that it contains in the way of 'theory' or pseudo-science, which includes a whole area of 'technical knowledge'" ([1973] 1987, 142).

To the contrary, Freire says that the agronomist while accepting the life, culture, and beliefs of the peasant farmer, must challenge the farmer to think about why he uses the techniques that he does. The agronomist, however, cannot approach dialogue as though he confidently existed within a closed circle of knowledge. If he does, then he suffers from an "absence of doubt" ([1970] 1988, 23). He must be open to the possibility that the peasant's knowledge and technology may be more appropriate than his own within the existing context. The expert must be open to the possibility of ignorance within himself.

In other words, the expert must enter into real dialogue with the peasant — a dialogue in which the expert's perceived expertise does not preclude any final decisions. Pacey, who has much in common with Freire, shares this conviction about genuine dialogue as well. "To engage in a genuinely open dialogue," Pacey writes, "is inevitably to share power over the final decision" (1983, 157).

TRADITIONAL WISDOMS AND SCIENTIFIC VISIONS

Likewise, Goulet expresses what many technical experts have learned firsthand: that they need the knowledge of the peasant as much as the peasant needs theirs. The expert needs the peasant to validate the imparted technical knowledge within the peasant's own context. While the experts must challenge the assumptions and practices of the traditional wisdoms, "these wisdoms will need to criticize the value premises of the scientific visions" as well ([1977] 1989, 240).

Eugene B. Shultz, Jr., a professor of chemical engineering who has done research in Central America and in Africa on the use of root fuels for cooking, offered me a good example from his own experience of how this reciprocal relationship works. One of the major problems in the Third World is the constant depletion of forests and the lack of available wood for fuel. Shultz has experimented with roots from the southwest United States that can be grown by the peasants in their own environments and used as a fuel substitute for wood. However, to test the viability of the idea as well as the efficiency and acceptability of the various root types, he had to enlist the cooperation and knowledge of local women.

Many factors beyond simple combustion are important in the choice of fuel for the cooking of a meal. The fuel might convey an unpleasant odor. Or it might fail to provide a buildup of soot that is used as a form of weatherproofing in the peasant home. Only local women, actually involved in preparing a meal, could attest whether the root fuel was more acceptable in their environment than their traditional fuels. Their knowledge was not solicited out of politeness or diplomacy. It was essential to the entire scientific experiment. This is a clear example of what Goulet calls for — traditional wisdom criticizing the scientific vision.

Freire's thinking is also along these same lines. The technical expert, according to Freire, must attempt to get the peasants themselves to examine their own experiential knowledge — about planting, erosion, animal breeding, herbal medicines — because this knowledge is permeated with peasant attitudes about religion, about ancestor cults and cults of the dead, and many other nonscientific beliefs. When the expert attempts to modify traditional techniques that are governed by belief as well as experience, however, he threatens both local beliefs and knowledge. The expert must realize that the knowledge and the techniques he is offering as a substitute are also socially and historically conditioned ([1973] 1987, 108).

PARADOX OF TECHNOLOGY

It would be a mistake, though, to think that Western thought and technology have nothing to offer the Third World. That is certainly not where this discussion has been heading. There is a sense in which Western technology has estranged Western humanity from nature and from its own community. It has precipitated isolation, violence, militarism, and many other forms of fragmentation within the community and the human psyche. But, paradoxically, this technological environment has also produced new explorations of the human interior. Walter Ong, a scholar in both literature and contemporary civilization, writes of this in several of his books — and in a hopeful spirit. Ong sees clearly that technology itself has made it possible for us to explore the earth and the solar system, to bring even scattered and remote human communities to our attention and consciousness, and to study the human brain and the human person in ways that were previously impossible (1967, 133).

Although the Western nations have become more aware of other communities throughout the earth, they have yet to fully develop an ongoing sense of responsibility toward the poorest of those communities and a global ethic that will define those responsibilities beyond the level of intermittent aid for famine victims in Africa or earthquake victims in China. Most have yet to really see the issue of Third World development as one of justice and love. These are concepts that have occupied philosophers throughout history. Plato, for example, wrote of justice. John Rawls is still writing about it today. In a real sense, though, justice and love are prephilosophical concepts. When you see for yourself hundreds of homeless persons living on a garbage dump in Port-au-Prince, Haiti, washing their clothes in an open sewer, and drinking the same water, you do not need to be told what injustice is. You are staring at it, and your very humanity, unless it is pathological, tells you it is wrong. Love is the response to correct that wrong.

Western technology and expertise, however, must not be thrust upon the Third World even in the name of justice and love. Development cannot come after the turbines are turning and the expert systems are outputting their diagnoses and decisions. It must come at the beginning. The decisions to build the turbines and design the expert systems are part of the development process. This process entails making the decision for change and exercising one's own rationality to achieve that change.

Clearly, the exercise of expertise involves decisions and choices. Expertise in engineering might be applied to the construction of Tomahawk missiles or artificial hearts. Expertise is, in effect, a normative concept that involves evaluations of what will work and what will not, what should be done and what should not be done. As such, the definition of expertise is intimately linked to cultural and ethical values. The introduction of expert systems into a Third World environment, consequently, could be intrusive and create problems rather than solving them.

Many will argue that it is impossible to talk to Third World villagers about sophisticated expert systems and microchips. These technical details are beyond their comprehension. That may be true. But planting a crop is not beyond their comprehension. Healing a wound is not beyond their comprehension. The knowledge for an expert system for agricultural planning or for medical diagnoses begins with the experiences of planting and healing.

If expert systems are introduced into the Third World as though they are simply Western know-how on a microchip, then they will not weaken Third World bonds of dependency upon the West but strengthen them. Local expertise about agriculture or medicine, for example, may be dismissed because the knowledge comes from a farmer not an agronomist, or from a traditional healer not a physician.

Even in the West, as knowledge becomes more specialized and more "technologized," individuals lose confidence in their ability to think and act on their own behalf. They are quick to turn to doctors, mechanics, plumbers, termite specialists, TV meteorologists, and countless other specialized sources of knowledge rather than attempt "home remedies" or do-it-yourself projects.

In many Third World settings, however, you cannot be a "specialist." You must know your environment simply to make the decisions that allow you to survive in it. Moreover, this knowledge defines the individual personally, socially, and, often ethnically. Expert systems, on the other hand, not only incorporate knowledge, they, in effect, alter our concept of what knowledge is. If expert systems are looked upon as a better replacement for local knowledge and decision, then they become a threat rather than a tool for development.

Chapter 5

Knowledge, Wisdom, and Computers Cum Laude

> To know that you know what you know and that you do not know what you do not know — that is knowledge.
>
> — Confucius

> Knowledge—the small part of ignorance that we arrange and classify.
>
> — Ambrose Bierce

> Addiction to knowledge is like any other addiction; it offers an escape from the fear of emptiness, of loneliness, of frustration, the fear of being nothing.
>
> — Jiddu Krishnamurti

The expertise that expert systems theoretically incorporate is sometimes characterized as meta-knowledge — knowledge about knowledge. This expression is itself problematic. The history of epistemology since the early Greek philosophers has largely been an ongoing attempt to arrive at just that: knowledge about knowledge. What is knowledge? How do we know? How do we know that we know? Can we acquire knowledge other than through our senses? What are these representations of the world that the mind reflects and acts upon? How is it possible to have knowledge of time and space that we have not immediately experienced? The distant past? The future? The exotic? What are the limits of knowledge? What are the limits of communicating knowledge between individuals and across cultures?

With the emergence of a computer technology that, perhaps unfortunately, got tagged by enthusiasts in the laboratory as well as in the advertising office as Artificial Intelligence, cognitive scientists have turned their attention to this new kind of knowledge. There is now a sizable literature of debate on whether computers are intelligent or whether we are wasting time and money in trying to make silicon chips perform like the human brain. It is not my intent to review the work of those writers, such as John Haugeland, Hubert Dreyfus, Allen Newell, Herbert Simon, Daniel Dennett, Marvin Minsky, John Searle, Zenon Pylyshyn, Paul Churchland, Jerry Fodor, Terry Winograd, and many others. Their names and arguments are now well known to anyone interested in both computational models and connectionist models of thought.

SPLIT BRAINS AND NARROW MINDS

Some of the most interesting research on human knowing is now coming from the area of cognitive neuroscience. The new collaboration between neuroscientists, cognitive psychologists, computer scientists, and philosophers of mind will certainly bring a different orientation to AI research (Lister & Weingartner 1991; Gluck & Rumelhart 1990). Although not directly linked to these particular interests, Oliver Sacks, a British clinical neurologist, wrote a fascinating book in 1985 called *The Man Who Mistook His Wife for a Hat.* The "clinical tales" that Sacks relays are about neurological losses — losses of memory, losses of identity, losses of particular functions — and about the extraordinary effects these losses have upon individuals' lives. Like A. R. Luria, who saw the limitations of an impersonal science, Sacks deliberately chose to use a story format because he sensed that a case history was designed only to describe the disease, the "what," but a tale could convey something of the history of the person experiencing the disease, the "who." Others in psychology have followed suit, such as Susan Baur with *The Dinosaur Man: Tales of Madness and Enchantment from the Back Ward.*

One man that Sacks describes had damage to the right hemisphere of his brain. Sacks reiterates the traditional distinction that the left hemisphere of the brain, in terms of human evolution, is more specialized and has to do with abstraction and categorization, with language, mathematics and logical thinking, with symbols and concepts. But the right side of the brain has to do with, what Sacks calls, recognizing reality — something which every creature needs to survive. The right side of the brain, in Sacks's description, is more holistic. Specific objects and events are more meaningful to it than abstractions. The ability to identify familiar faces is a function of the right side of the brain. Also, interestingly, is the ability to understand jokes, as well as figures of speech and subtle shades of meaning, such as metaphors and irony. There is clear evidence for this, some of which is also discussed by Stephen Kosslyn and Olivier Koenig in *Wet Mind*. In one experiment, when patients with damage to the right hemisphere of their brains were asked to select a picture that corresponded to the statement, "Sometimes you have to give someone a hand," they chose a drawing of someone offering another person a hand on a tray (Kosslyn & Koenig 1992, 245). Their undamaged left hemispheres could grasp the literal meaning of the sentence, but not the metaphorical intent.

Many of these characterizations about regions and functions of the brain, of course, have been the result of studies on patients who have sustained brain damage. If a patient, for example, exhibited a language deficit following an injury to the brain, then it was concluded that the damaged area of the brain must be the "site" of language. With new tools, such as positron emission tomography (PET) scanning, there are other ways to map brain activity. Some of the simple functions of the brain may be localized in specific areas, but it is likely that the more complex functions, such as reading, are integrative and carried out by simpler functions that may be localized in different regions of the brain (Kosslyn & Koenig 1992, 11–14).

In fact, recent research on split-brain patients indicates that this traditional identification of the left hemisphere of the brain with language and the right hemisphere with imagery is an oversimplification in the case of mental imagery. Visual imagery is the result of sensory input. Mental imagery, on the other hand, is seeing in the absence of sensory input — seeing "with the mind's eye." Generating a mental image apparently involves at least two processes — one that activates stored shapes and one that uses stored spatial relations to arrange the shapes into a mental image (Kosslyn 1988, 1626). The left hemisphere seems to be better at arranging shapes with categorical information, while the right hemisphere is better when coordinate information is required (Kosslyn 1988, 1626). So it is likely that neither hemisphere alone is the source of mental imagery.

In any event, Sacks's patient, with his damaged right side of the brain, had completely lost the concrete, the personal, the whole, the real in his life, and he had been reduced simply to the abstract, the detailed, and the categorical. The title of the book comes from the fact that in leaving Sacks's office the man reached out his hand, took his wife's head, and tried to lift it off. He had apparently mistaken his wife for a hat. And Sacks says that "his wife looked as if she was used to such things"

(1985, 10). This poor fellow could not visualize faces (even those of his family) or familiar scenes; he could not identify common things, such as a shoe or a rose. Thinking that they were the heads of little children, he would casually pat the tops of fire hydrants. He was incapable, in other words, of perceiving identity or particulars, and this made him incapable of human judgment. But, interestingly enough, he was fine with the abstract. He could easily visualize schemata. He had no difficulty, for example, in playing a tough game of chess.

The disease the man was suffering from was called agnosia. It sometimes sounds as though agnosia may be what our AI systems are suffering from. We have computer brains with only left hemispheres. They do fine with computation and abstraction but not with the whole, the personal, the particular, the real. AI is limited not by the technical but by the social — the social being both the complexity of the environments in which the computer system is introduced and the complexity of understanding what it is we are really trying to do in those environments. In reflecting on the introduction of expert systems to the Third World, my primary concern is not epistemology per se, nor is it theories of cognition or theories of brain function. It is understanding the complexity of the task before us and the complexity of the technology.

THE KNOWLEDGE INFRASTRUCTURE

In recent years, a new emphasis has been placed on the need for the nations of the South to "regain the capacity to generate scientific and technological knowledge" and not simply assimilate imported technologies and knowledge (South 1990, 16). Third World nations face a dilemma, though, in their attempts to develop indigenous knowledge and technology. In *The Knowledge Context*, Philip Altbach examines the inequality that exists between First World and Third World in the system for dissemination of knowledge, which, not surprisingly, Altbach says, is "dominated and controlled by the large, wealthy, and prestigious academic systems, publishers, and research enterprises in the major industrialized nations" (1987, xv). Although there are some notable exceptions, such as India, Altbach maintains that "it is clear that the Third World relies on the West for most of its knowledge distribution" (1987, 184). In his study on Peter Ramus, an educational reformist of the sixteenth century, Walter Ong notes that a passion for textbooks and curricula has been one of the Western world's remarkable peculiarities and has created its unique role:

All other cultures — which have their own particular and different contributions to mankind and to history — have, since the sixteenth and seventeenth centuries, looked to the West to improve their own techniques of propagating and assimilating knowledge. The West has given to the world not only universities, but the interests and techniques which go with the university state of mind. ([1958] 1983, 164)

In addition, language plays an important role in the knowledge network. English is the dominant language for the publication of scholarly articles in science and technology. French, Spanish, and Russian also have a large and international audience (Altbach 1987, xvi). This, in itself, clearly places Third World nations at a disadvantage. A Sengalese physicist would not publish his research in Wolof. To begin with, he probably would not be able to express certain concepts of modern physics in that language. Paraphrasing the American philosopher, W.V.O. Quine, how would it be possible to translate "neutrinos lack mass" into Wolof? In such cases, Quine, like Wittgenstein, maintains that "understanding a sentence means understanding a language" (1960, 76–77). But even if the physicist's journal article could be written in Wolof, it would go unnoticed because, within the community of scientists, only a handful would be able to read it.

To access current scientific and technological research and to contribute to it demands, today, a facility in the primary languages of the West. The major publishers, journals, data bases, universities, and research institutions that make a knowledge network possible and that provide the critical mass for intellectual spark and synergy are concentrated in the Western industrialized nations.

Surrounded as we are by universities and colleges, by a free and public system of libraries, and by a wide selection of bookstores, we fail to appreciate what ready access to books really means. A study, carried out in the 1970s under the auspices of UNESCO, documented the lack of books in the Third World. Altbach quotes from the statistics in this survey that there is a shortage of books for 70 percent of the world and that the industrialized nations, with only 30 percent of the earth's population, produce 81 percent of the world's books (1987, 18). Even though the amount of publication in some Third World areas, such as Southeast Asia, has improved in the intervening years, Altbach notes that "the general configuration of inequality and shortage remains" (1987, 18).

The Third World nations are importing not just tractors and F15s and VCRs, but textbooks, journals, and knowledge from the West in a one-sided trading arrangement. In some cases, even importation and translation of materials is made more difficult with copyright laws, although complete enforcement of these laws is impossible, especially in foreign nations. "Perhaps most important," Altbach suggests, "books published in the industrialized nations are written, edited, and published for an audience in these countries and not to meet the needs of the Third World" (1987, 34).

VALIDATING INDIGENOUS KNOW-HOW

The challenge of promoting and validating indigenous knowledge for use in locally developed expert systems is complicated, however, by more than just the Western monopoly on the knowledge infrastructure. There is also the fact that many Third World nations have very diverse languages and ethnic subcultures within their own borders. In the case of the African nations, for example, this is largely the result of the way in which colonial territories were carved out and negotiated by the political powerbrokers of nineteenth century Europe. Nigeria certainly stands as an extreme example of cultural and linguistic diversity in the Third World. There are approximately 250 ethnic and linguistic groups in Nigeria with languages that are mutually unintelligible. The three major languages, Hausa, Yoruba, and Igbo, are used by only about one-third of the population.

Throughout history, those in power — from without and from within — have used language as a means of control, suppression, and exclusion of ethnic groups. In some cases, Third World elites have retained colonial languages in government and administration for just that reason. Without a doubt, future computer development in Third World nations will mainly be in the large cities where the computer hardware and the personnel resources of corporations and universities are concentrated and where there is a plug into the knowledge network of the West. This will inevitably have an effect on what types of expert systems are developed.

The planning, construction, and organization that economists and technologists generally label development in the Third World have often been development for the cities at the expense of the rural areas, development for industry at the expense of agriculture, and development for the educated at the expense of the illiterate poor. If Third World decision makers and development workers do not clearly understand this new technology, expert systems could easily conform to these same biases. Which language will be used for the conception and design of an expert system, who will design it, whose expertise will be encapsulated, to what use will the expert system be put — all these questions become particularly significant in the Third World setting.

Even if the individual expert systems are designed locally within the Third World nation, the technology itself is still, in a sense, imported since it was conceived in Western universities and laboratories. All technology, in some way, reorganizes time, work, social attitudes, relationships, and authority around the cultural values of the environment in which the technology was developed. In doing so, it sometimes fragments the community. Anthropologists have documented all this firsthand.

A generation of development workers and Peace Corps volunteers are also well aware of this. If you introduce a water pipeline and handpump to a village, you have altered village life. Now, the village women — who are generally the ones to go for water and carry it back to their homes for cooking and cleaning — must learn and accept social practices that go with the handpump, such as waiting in line to take one's turn. Someone in the village must also take the responsibility to learn about the maintenance of the pump and fix it when it becomes broken. Others in the village may be forced, for the sake of new standards of hygiene, to look for an alternate place to water their animals, perhaps at great inconvenience and with great protest. And so the changes reverberate through the daily life of the village — or the new technology is abandoned.

INFORMAL KNOWLEDGE NETWORKS

It is also obvious that some of the knowledge to be found in Third World nations, where illiteracy rates can rise as high as 80 percent or more, will not be the same kind of knowledge that Western nations disseminate in their classrooms and store on their computer data bases. The scholarship of Eric Havelock, Walter Ong, Jack Goody, and others has shown that in oral and residually oral cultures, knowledge is often communicated in ways that encourage memorization and remembrance. These formulaic methods might be proverbs, songs, rituals, or plays.

In an oral culture, knowledge is dependent upon existent human memory. Knowledge is incorporated into formulas that will facilitate that memory. What is important is not the "news of the day" and the "latest development." What is important is what has always been. The critical point is incorporation, not information. While there is this formal element to the expression of the knowledge in a residually oral culture, the knowledge itself is informal in the sense that it is not delineated in a methodical, analytical way.

A computer, by contrast, is a very formal machine. John Haugeland has referred to the computer as a formal system because it is completely self-contained, its rules and symbols are definite, and it is finitely checkable ([1981] 1987, 6–10). You can say what the state of the computer system is at any given time. In order to have "time bytes" and "space bytes," you must digitize both time and space. But digital time is not time itself. For most of human existence, time was not measured; time was experienced in the changes of our natural and interior environments. We became timekeepers late in our evolutionary journey. In digitizing time and space, we have traded actuality for increased manipulation.

In the last decade, there has also been a new emphasis in development studies on informal politics and informal economics in the Third World nations. The distinction between formal and informal sometimes becomes blurred. However, according to the literature, formal political organizations derive their authority from the state and from legal institutions (March & Taqqu 1986, 1, 4). Informal political associations also have power and legitimacy but these are derived from other sources, such as kinship relationships and ethnic relationships. In addition, the boundaries, membership, and hierarchies within these informal associations are less distinct (1986, 9). What is true of political associations is also true of economic associations. The informal economy, for example, could be anything from the black market to the urban poor who gather cans or papers for reimbursement from recyclers.

Informal knowledge networks must somehow underlie all of this activity that is unmonitored and, therefore, invisible to the statisticians. I am not aware of any research on how knowledge of these informal arrangements and associations is communicated. Admittedly, such a study would be difficult to do. What exactly would you be looking for? I realize that, in an effort to conceptualize, we often use inflated terminology — like knowledge networks — to describe what is just the day-to-day exchanges of poor people trying to sell their melons, buy a goat, borrow some oxen, store some wheat, and find out what those with money are willing to buy and those with power are willing to sell.

We look upon all this as such obvious behavior that it requires no explanation. Scientists, however, and especially AI researchers, are always amazed by how little we know of the obvious. How do we recognize a face or a voice? How do we carry on a conversation?

PROVERBS

While we still know very little about informal networks for the dissemination of knowledge, there have been studies on some other methods for communicating knowledge in Third World cultures, particularly proverbs. "Use your tongue to count your teeth." "Remove the hand of the monkey from the stew or else it looks like the hand of a human." "If a chicken pecks and misses, his neck gets twisted." "Water may cover the footprint on the ground but it does not cover the words of the mouth." "Wherever you run, there is nowhere you don't touch the ground." These are just some of the Igbo proverbs that Joyce Penfield examines in her book, *Communicating with Quotes* (1983, 111–20).

While most of us would definitely be challenged to come up with any meaning at all for these sayings, Penfield points out that such proverbs are a prevalent and authoritative form of speech, not only in Igbo society but throughout Africa (1983, 1). "Authoritativeness is bestowed on the quote," Penfield says, "by virtue of its association with the highly respected authorities of the community, the 'experts'" (1983, 8). The meaning of a proverb, according to Penfield, is directly tied to a context. It is intended to achieve an anticipated effect for a specific situation and audience, to apply certain social values and norms to the resolution of a conflict, and to reveal good judgment. Chinua Achebe's novel of the Igbo culture, *Things Fall Apart*, is knotted with proverbs, which Achebe calls the "palm-oil" with which words are eaten.

Certainly, we recognize that, in some sense, there are different kinds of knowledge. Knowledge of the self, knowledge of others, knowledge of the world, knowledge of techniques, knowledge of the artist, knowledge of the mystic. Many of these proverbs have to do with self-understanding and social interaction, with reducing conflict, engaging conversation, and even sparking amusement, rather than with practical, technical knowledge. Their prevalence even today gives us an important insight into a Third World culture. Still, how do you incorporate this on a computer system? By computerizing the proverbs or proverbalizing the computer system? That latter suggestion may not be as facetious as it sounds. One researcher, born in Nigeria and active in village schools, has suggested using traditional Annang proverbs in Nigerian curricula as a method of education, since they are more appropriate to a community rooted in oral culture and tradition than some of the Western materials aimed simply at imparting literacy (Essien 1978).

WISDOM: KNOWING ONE GREAT THING

Clearly, proverbs convey wisdom, not technical instructions. It is difficult to say what wisdom is. In the seventh century BCE, the Greek poet Archilochus wrote that the fox knows many things, but the hedgehog knows one great thing. Although some people feel that everything they need to know they learned not from kindergarten, but from their cat, it might be stretching it a bit to call a hedgehog wise. Still, there is a sense in which wisdom really is knowing the one great thing that can defeat even knowledge. Focused as we are on information and data, we seem to rarely give wisdom a thought in our high-tech society. Robert Sternberg, a psychologist at Yale University, recently edited a book of essays by some of his colleagues that documents, however, the growing interest in wisdom as a psychological construct, and not simply a philosophical or religious one.

Wisdom does not appear to be a different kind of knowledge but a different kind of knowing, an integrative approach that refuses to look aside from the real human situation of uncertainty and ultimate death. It is uncommon sense. When J. Robert Oppenheimer saw the first atomic bomb exploded near Alamogordo, New Mexico, at the Trinity test site, he joined in the team celebration that this engineering design, which had required the orchestration of hundreds of the most brilliant physicists, chemists, and engineers, had actually worked. After Hiroshima and Nagasaki, his exuberance faded. Within two years he would give a talk at MIT (Massachusetts Institute of Technology) and say that, with the production of the atomic bomb, physicists knew sin, and it was a knowledge they could not lose ("Day After" 1980). In still later interviews, he would recall that day at Trinity. "We knew the world would not be the same," Oppenheimer reflected with infinite sadness. "A few people laughed. A few people cried. Most people were silent. I remembered the line from the Hindu scripture, the Bhagavad Gita: 'Now I am become Death, the Destroyer of Worlds.' I suppose we all thought that — one way or another" ("Day After" 1980). The ability to construct the bomb was knowledge. The ability to see that the bomb should not have been constructed and certainly not used was wisdom.

In many Third World cultures, where formal education is still the exception, there is a greater emphasis upon wisdom than upon knowledge, a greater respect for the wise judge than for the scholar. All of this will have a clear impact on any effort to use indigenous knowledge in expert systems designed for the Third World environment. In the Third World context, the challenge will be to formalize for the computer a knowledge that exists only informally now — and to convince the traditional farmers, environmentalists, and healers that this is a wise thing to do.

THE LOCUS OF KNOWLEDGE

The other alternative, of course, is to import expert systems. To realize what is at stake in importing knowledge-based technology into the Third World requires at least some understanding of what knowledge really is and how knowledge that is symbolically represented in one language and culture is translated into another language and culture. If their intention is genuinely to empower the poor in their nations, Third World decision makers and strategists cannot avoid this deeper look into expert systems and the larger implications of its use.

The nature of knowledge and the relationship of knowledge to language and culture is, of course, a huge area for thought and research — and one which is open-ended. I am not implying that it can be tackled in a single chapter of a book. At a minimum, we need help from the fields of philosophy, psychology, anthropology, and linguistics when we begin talking about knowledge, language, and symbols. We need help to clarify our assumptions and beliefs, to sharpen our arguments and observations.

Still, general theories of philosophy and psychology, of anthropology and linguistics come and go. It is difficult to know when they are on their way in or on their way out — and difficult to predict which will have the most enduring insights for the task at hand. All knowledge necessarily advances with uncertainty and revision. Those who choose to write about any complex subject must begin by accepting human and personal limitations and realize that their job is not "to get it right" but simply to bring the best insights that they presently can to the issues.

As one would expect, our understanding of what knowledge and reason are has guided our design and implementation of expert systems in the Western nations. This has been a rather common-sense understanding. We took it for granted that we knew what knowledge was and what reason was. Knowledge was a growing collection of more-or-less discrete facts that were assembled in textbooks, encyclopedias, and dictionaries: the boiling point of water, the date of the Norman Conquest, the difference between an iamb and a trochee, the number of subatomic particles in each chemical element, and so on and on.

According to this popular belief about knowledge, first-class experts, such as physicians, held hundreds or thousands of important and specialized facts in their heads: forty-five is a normal hematocrit; excess glucose in the urine is usually due to diabetes mellitus; phenytoin is widely prescribed as an anticonvulsant. We also acknowledged, in theory, that there could be experts outside the universities and the professions, such as an expert in chess or an expert in fighting oil-well fires. An ad for Hitachi implies that expertise is not the sole province of the human species. Two cats are pictured intently watching a large television screen with an underwater setting of coral and tropical fish and a caption that reads: "Only Ultravision could fool these experts."

No matter how we came by them, all our facts of knowledge were assumed to be somehow asocial and acultural. The Civil War began with the Confederate bombardment of Fort Sumter in Charleston Harbor at dawn on April 12, 1861. Translate that statement into the language of the Khoisan (complete with its characteristic clicks), teach it to Khoisan children in Botswana, and they would then possess this piece of knowledge as surely as a seventh-grade student in St. Marys, Kansas. That, more of less, was the story we told and believed. The locus of knowledge, in other words, was the individual, not the society or the culture.

What was true of knowledge was also true of reason. Reason was an innate human faculty of transcendental validity. Very often in our journey toward fuller human consciousness, we have denied that humanity extended to those outside "our own." To the Greeks, non-Greeks were barbarians. To many of the New World "conquerors," the indigenous people of both Americas were savages to be either exterminated or impressed into slavery. There have been "killing fields" throughout human history. When we did come to recognize humanity under different colors and cultures, we still assumed that others must reason as we do — or be taught to do so. The rules of reasoning — rules that determined what conclusions you could draw from any existing set of facts — had to be universal to the Dzu Bushman, the Aleut, and the Oxford don.

In this scheme of things, computers seemed ready-made to replace the human brain as the locus of knowledge and reasoning. While the human brain can store an estimated 100 trillion bits of information, including neurophysiological information, it was assumed that more discrete facts could be stored in all those gigabytes and terabytes of computer memory than could be stored in a single human brain. And the rules of reasoning could be coded and programmed to allow the computer to draw conclusions, in the manner of a human expert, from all the bits of information. All this was the dream of the early researchers into Artificial Intelligence and expert systems.

As it turns out, we may have been too quick to assume that we understood what in fact we did not. A little knowledge about knowledge is also a dangerous thing. We do not really know yet how knowledge is retained in our own brains. Most of us have had the experience of trying unsuccessfully to come up with a name or a date from our memory, only to have it surface in our thoughts minutes or hours or even days later when we were not making a conscious effort to recall it. Cognitive psychology has largely been an attempt to devise lexical tasks that will help us to understand how information is stored and how memory is facilitated in the human brain.

In addition, we do not know exactly how new knowledge is extrapolated from our existing knowledge. Even science has its mysteries. In *The Search for Solutions*, Horace Judson, a respected science journalist, recounts many cases in which answers have come to scientists unexpectedly, in flashes, in slips of the tongue, even in daydreams. Friedrich Kekulé, the German chemist who discovered the structural basis of organic compounds, described how, in such a daydream before an evening fire, he saw rings and chains of carbon atoms dancing before his eyes (1980, 6–7). Here in twilight consciousness was the solution that had eluded his more aggressive efforts.

It is also uncertain how new knowledge is incorporated into existing knowledge — how it changes it or replaces it within our individual brains. Some new knowledge might simply be an isolated fact. We learn from a television documentary that Callisto is the oldest of Jupiter's satellites, and this is tucked away somewhere in our memory. More complex and coherent knowledge, such as the theory of evolution, must make sweeping alterations in what we already know and how we know it. How does all this knowledge in our heads get "rewritten" and why does some of it seem to persist even when it is proven to be contradictory? One thing is certain: we do not experience our own knowledge as though it were an internal textbook.

THE UNSAID

In the case of expert systems, there is the added factor of expertise. Just what is it? Expertise is supposedly extracted from all the other knowledge that the expert has and bound into a "domain" of knowledge. This assumes that knowledge is somehow compartmentalized within the human brain and that one area of knowledge can be circumscribed and excised from the human interior and incorporated in an external medium, such as a textbook or a silicon chip.

I have had the experience of working with many experts, particularly pathologists and physicians, in the design of computer software. Invariably, when you ask one of these individuals what it is they do, there is a great deal of head scratching, uhm's and ah's and well's, starts and backtracking. Never have they sounded as though they were reciting *Gray's Anatomy* or a manual on the clinical interpretation of laboratory tests. In fact, they would often hand you one of these manuals and say, "This is what we do here," even though a few hours in the laboratory and some interviews with their laboratory technicians was enough to confirm that it was not really what they did — exactly. To have put the information from the textbook manual directly into a computer system would have been, at best, unsatisfactory. Philip Johnson-Laird, a British cognitive psychologist, submits that, if nothing else, expert systems differ from their human counterparts because the latter are better at making excuses when they are wrong (1988, 226).

These professional experiences of mine and of many others in the computer field should raise questions in our minds about what it is that expert systems actually do. Can one area of knowledge really be isolated and extracted from the whole noetic process that takes place within a human person? Can we ever fully articulate knowledge? Stephen Tyler grapples with this latter question in *The Said and the Unsaid.* The said consists of the saying itself, the reconstruction of what was said, and the anticipation of what will be said (1978, 459). The unsaid consists of all the unspoken presuppositions and implications that are background to the saying, both "those created conventionally by the 'said' and those created intentionally by the speaker and hearer" (1978, 459).

Paraphrasing Tyler, we could ask if something is always left unsaid, how do we interpret what is said — and what is not? And, is knowledge really something that can be isolated, transmitted, received, stored, translated, and programmed, or is it an ongoing process within an individual human person that eludes and defies these attempts at cognitive description? Is what is knowledge in one culture necessarily knowledge in another?

KNOWLEDGE AND INFORMATION

It is probably not incidental that, with expert systems, we ceased to talk about data bases and began to talk about knowledge bases. There is a subtle recognition that knowledge is not just data or bits of information, even though, in ordinary conversation, the terms information and knowledge are sometimes used interchangeably. The *Dictionary of Computing*

(1990) defines information as "collections of symbols." It is equivalent to "data." A data item might consist, for instance, of a certain number of "information bits" that are capable of conveying "meaning" and other bits that are simply for error detection.

Symbols are capable of transmitting meaning, but whether meaning or the intended meaning is actually transmitted is dependent upon the recipient of the transmission. If I pick up the Torah, there is meaningful information there; but unless I can read Hebrew, the symbols on the pages are not meaningful to me apart from being recognizable, perhaps, as linguistic symbols. Or in certain circumstances, I may see meaning in symbols but it is not the intended meaning. Human beings do not always say what they mean or mean what they say. At times, they do not even know what they mean — or know what they have said. With a wink to my philosopher friends, I recall that it was Voltaire who wisecracked through one of his fictional characters in *Candide*, "Quand un homme parle à un autre qui ne le comprend pas et que celui qui parle ne comprend pas, c'est de la métaphysique," which is sometimes translated as "when he who hears does not know what he who speaks means, and when he who speaks does not know what he himself means — that is metaphysics."

Then there are the instances when we say even more than we know. Murray Gell-Mann, a physicist who shared in the 1969 Nobel Prize for his work on subatomic particles, explained how he once came to an important insight when he misspoke. He was attempting to describe "strangeness" — one of those playful names, including charm and quark, that physicists have given to subatomic particles and their behavior. Instead of saying "isotopic spin — three halves," he said "isotopic spin — one" and what was unthinkable suddenly became not only thinkable, but correct (Judson 1980, 21).

Information can be carried on light, sound, or radio waves; on electric current; in magnetic fields; through marks on paper or stone or human flesh — conceivably on any physical material. It does not always originate with human persons — animals, for example, communicate information — and it is not always directly observable. The human genome, with all our genetic information encoded and coiled in DNA, contains an estimated three billion nucleotides, of which perhaps 70 million have already been parsed. There is an ambitious project under way to create genome data bases — to translate this biological information into digital information and store it on computers so that it can be "accessed, integrated, queried, interpreted, visualized, tested and studied" (Erikson 1992, 128).

Fred Dretske defines information as "a commodity that, given the right recipient, is capable of yielding knowledge" (1981, 47). Unlike knowledge, information is a commodity — it can be transmitted, received, stored, translated, exchanged, lost, restored, corrupted, truncated, bought, sold. And "given the right recipient," it is capable of yielding knowledge. At an intersection, for example, there are two gas stations. The sign outside the Shell station says "Gas — $1.20 a gallon." The sign outside the British Petroleum station says "Petrol — $0.30 a litre." If I know (a) that petrol is gas, (b) that a litre is larger than a quart, and (c) that there are four quarts in a gallon, then I should also know that the gasoline is cheaper at the British Petroleum station than at the Shell station.

The amount of knowledge that information yields, therefore, is dependent not just on the recipient's ability to recognize the information as information, but also on how much previous knowledge the recipient has. Information is, in some sense, external to the subject; knowledge is internal. Information is incorporated into knowledge within a human person. Knowledge is interpretation of the information. Even sensory perception is a matter of interpretation. The brain does not simply take in sensory input; it interprets that input. Clinical studies of optical illusions, for instance, have made this increasingly clear to us.

RULES OF OPPOSABLE THUMBS

In the specific case of experts, moreover, we have learned that the rules by which an expert incorporates and interprets information and then makes a judgment based upon his knowledge are never wholly explicit — even to himself. Knowledge can never fully be extricated from the human interior in which it resonates.

Amos Tversky and Daniel Kahneman are well known for their research into the nature of decisions. They have shown that certain heuristics and biases are concealed in the decision process. A heuristic is a guide or rule of thumb that is used for assessment and prediction in making a judgment. Tversky and Kahneman demonstrate that the rule itself may be unproven or incapable of proof; it may also lead to systematic error. Some probabilities, for example, are evaluated by the degree to which *A* is representative of *B*. But such an approach is insensitive to prior probabilities, sample size, and predictability. Also, different starting points tend to yield different estimates — a phenomenon referred to as "anchoring" (Tversky & Kahneman 1974, 1124).

Clearly, the heuristics that guide the decisions of the illiterate farmer in Kenya will not be those that guide the farmer in Kansas with a university degree in agriculture. Does this fact in itself exclude the Kenyan's "expertise" from consideration? Does it exclude his participation in the design of an agricultural expert system for his community? Should it?

The underlying argument here is that culture impacts what we know and how we go about knowing it. The encoding of experience, categorization, inference, problem solving, and reasoning are not only cognitive processes, but cultural processes as well. As a result, we can easily confuse cultural bias with absolute, value-free "knowledge."

KNOWLEDGE IMPLOSION

When we propose, therefore, to symbolically represent expertise for the computer, we are attempting to encode some partial description of "The World" that we have perceived and constructed through our own language and our own culture. This encoded information may not be a reliable basis for decision in another culture, another context. This is not cultural relativism but cultural realism.

Moreover, expert systems attempt to simulate and, in some cases, uncover the reasoning process. This is technology on a much different level than instrumental, mechanical, chemical, or biological technology. It comes close to touching who we are and how we think. With biotechnology, such as gene splicing and in vitro fertilization, we are clearly manipulating even the human life process in a very intimate way. But it is still our communal knowledge, not our individual genetic makeup, that defines our humanity. In fact, we are told that we share with the chimpanzee approximately 99 percent of our genetic material (Wilson 1984, 130). Therefore, strategies to transport expert system technology from Western nations to the Third World must address epistemological issues — issues that deal with the limits and validity of knowledge.

This includes the limits and validity of our own Western knowledge. The growth of information about our world has accelerated to the point that it can no longer be stored simply inside the human cranium. It is stockpiled outside the brain in computer data bases and in libraries with book titles that run into the millions. But not all of this human-store represents "knowledge." Some of it is redundant; some of it has, with time, been proved inaccurate; and some of it always was untrue. Each of us knows — absolutely, positively — things which are not so. It is difficult to measure the "knowledge explosion" that we so often allude to.

I am more concerned about the knowledge implosion that takes place within the human mind.

Certainly, there are whole fields of knowledge that did not even exist fifty years ago — and new technologies that have sprung from this knowledge. But often our technology conceals our lack of knowledge. In the United States, medicine, for instance, has become a very technological field, but we often lack real evidence about what works and what does not when we turn to all that waiting technology. More important than what we know or what we do not know is what we want to know.

Chapter 6

A Cage for Reason: The Formalities of Knowledge

Science is an attempt to make the chaotic diversity of our sense experience correspond to a logically uniform system of thought.

— Albert Einstein

There are children playing in the street who could solve some of my top problems in physics, because they have modes of sensory perception that I lost long ago.

— J. Robert Oppenheimer

In the expert system, the rule becomes the formalism of knowledge. The contention is that once the expert clearly states his rules and procedures for us, we have the expertise virtually tucked away in our pocket. Expertise, from this perspective, is something that can be formalized or expressed in axioms. The whole inference process in the expert system suggests that knowledge can be set out in facts from which conclusions and new facts can be drawn.

Since the time of Aristotle, we have entertained the idea that knowledge can be formalized and categorized and axiomatized. Walter Ong suggests in *Ramus, Method, and the Decay of Dialogue* that this Western compulsion for formalization was intensified in the sixteenth century through the influence of a Paris professor, Peter Ramus (Pierre de la Ramée), who, ironically, attacked Aristotle, or at least the medieval interpretation of Aristotle ([1958] 1983, 172–75). In some early printed treatments, the "explanations" of the logical method remain iconographic — they are pictures. The *Mnemonic Logic* of Thomas Murner

(1475–1537), for example, is about one-third illustrations in which, Ong explains, "various terms and rules are associated with various designs in ways calculated to beat the terms and rules into little boys' heads if not really to teach any genuinely formal logic" ([1958] 1983, 85). A bell, an acorn, a lobster, a cat, a scorpion, a fish, a bird — these are just a few of the designs in Murner's book that stand for logical terms, such as, an enunciation and a syllogism (Ong [1958] 1983, 85). Contrary to this medieval penchant for iconography as an educational tool, Ramus, according to Ong, produced the model of the "textbook" in which all subjects were approached, not through illustrative pictures, but through definitions, purportedly totally formalized, and through schematic tables and diagrams.

INNER ICONOCLASM

Ramism produced what Frances Yates has called an "inner iconoclasm" (1966, 235). The Ramist reform called for smashing the "inner image" as a mnemonic device and replacing it with the "imageless Ramist epitome . . . memorised from the printed page" (1966, 234). Following a trend for quantification that grew out of medieval logic, the Ramist method set forth verbal subjects and analyzed them into finer and finer subdivisions. Where previously knowledge had been enunciated, it was now presented through the medium of the printed, alphabetized word. To a great degree, the printed curriculum was considered the primary access to reality.

Ong has argued that this new methodology grew out of a "pedagogical necessity" ([1958] 1983, 23). As a new teaching master with the task of instructing, in Latin, large classes of squirming teenage boys about logic and rhetoric, Ramus, like generations of teachers after him, opted for a method that led to increased simplification of knowledge (Ong [1958] 1983, 39). Method became more important than the social context of dialogue.

Instead, Ramus presented knowledge "in terms of spatial models" that were intended to be grasped visually (Ong [1958] 1983, 9). Typically, dissecting a subject into progressive dichotomies, Ramus produced diagrams that resembled binary trees. With the impetus of the new technology of print — ready made to mass produce these tables and diagrams that were very difficult to reproduce in quantity by pen and ink — knowledge became an abstract, depersonalized, and presumably totally formalized object of the mind.

René Descartes, who was to lay the foundations of modern philosophy, learned his logic in a Jesuit classroom from a textbook that incorporated Ramus's concern with "method" (Ong [1958] 1983, 125, 198, 307). For Descartes, thinking was an operation upon symbols; thinking followed rules. Descartes's model for the mind was the science of geometry. Ong argues that it was Ramus's methodology, augmented by the print technology, that finally shifted Western society to an analytical, "scientific" way of thinking. Ramus's logic or dialectic was, according to Ong, a "residual logic" whose importance to intellectual history was not abstract theorems, but a set of "mental habits" that encouraged a spatial, visual orientation to thought and "seemed to signal a reorganization of the whole of knowledge" ([1958] 1983, 7, 8, vii).

QUANTIFICATION OF KNOWLEDGE

Since Ramus and Descartes, logic has become increasingly quantified and involved in the manipulation of symbols. In the twentieth century, Alfred North Whitehead and Bertrand Russell spearheaded the merger of mathematics and logic. From this came propositional, first-order, and higher-order logics in which words seemingly disappear entirely — although even the most abstruse, mathematical expressions are grounded in explanations that require words. The natural progression in this process of quantification has led to the computer in which knowledge, theoretically, is reduced to a binary code of 1s and 0s.

Certainly, there is simplification in reducing the knowledge of the expert to the textbook and then further simplification in reducing the textbook to the program code of an expert system. But when you distill a textbook to 300 or 500 coded rules, something has to have been lost. What was it? Why was the explication that was required in the textbook, not required in the expert system, and why is the explication in the textbook not exactly as the discipline is practiced — from one practitioner or "expert" to another? The formalization process has to leave something out and so the expert system has to be something less than the expertise of the expert. Looked at historically, it would appear that our formalist approach to knowledge is actually a cultural bias that needs to be recognized as such when we talk about expert systems and the transporting of these systems to other cultures.

Philip Leith has recently taken up some of Ong's insights about Ramus and the quantification of knowledge and used them to take a new look at the limitations of computer methodology, including expert systems. Leith argues that "it is impossible to set out our knowledge in the axiomatic manner which was deemed essential by Aristotle, for knowledge is not a simple grouping of isolated facts and empirical evidence as suggested by the syllogistic process" (1990, 22).

The central belief of modern science has been that reality existed "out there" and that facts of knowledge could be arrived at by performing experiments upon that reality. Out of this rationalistic tradition, came the assumption that expert systems can comprise rules of formalized knowledge. Leith and others do not buy into this rationalistic, empirical view of science. They propose, instead, a social theory of knowledge, which maintains that there is a tacit and cultural dimension to knowledge and expertise that can never be fully extracted into a rule or a scientific fact (1990, 30).

For Leith, there is a commonality between the methods of computer science and the methods of Ramus that Ong painstakingly documented. While Ramus used the technology of printing, programmers use the technology of computing — and to the extent that they both are formalist attempts at solving problems, they ignore the social context in which dialogue and expertise are embedded (1990, 80).

SOCIOLOGY OF KNOWLEDGE

The concern for what communities know and how they know it is also taken up by Harry Collins in *Artificial Experts*. As a sociologist of knowledge, Collins argues that scientific activity is inherently social and that "conclusions to scientific debates, which tell us what may be seen and what may not be seen when we next look at the world, are matters of social consensus" (1990, 4). He contrasts two models of learning or seeing. In the formal model of seeing, an object is detected by distinctive characteristics, whereas in the enculturational model of seeing, the same object may be seen as different things depending on the culture of the observer. "Thus," Collins states, "learning scientific knowledge, changing scientific knowledge, establishing scientific knowledge, and maintaining scientific knowledge are all irremediably shot through with the social" (1990, 5).

Since we assumed that the individual was the locus of knowledge — the one who observed, the one who deduced, the one who stored the facts of knowledge — we thought it quite feasible to transfer the knowledge from the individual brain to the individual computer. It was a simple matter of storing knowledge at the computer's memory locations in the form of bits of information and rules. All very well, except, if the sociologists of knowledge are correct, knowledge, like language, actually resides in the group, not in the individual. "Contrary to the usual reductionist model of the social sciences," Collins writes, "it is the individual who is made of social groups" (1990, 6).

There is clearly knowledge that is private. Gerard Manley Hopkins wrote in 1880 about the feeling of self that is incommunicable, even to those with whom we are most intimate. He speaks of "that taste of myself . . . more distinctive than the taste of ale or alum" (1959, 123). We each have this immediate sense of an unnamed self. Moreover, new "discoveries" are added each day to our scientific knowledge. While one or more names may be associated with these discoveries in the textbooks, they are often come upon almost simultaneously by different researchers. James Watson and Francis Crick were in a race to be the first to offer a model for the structure of DNA (the double helix) that accounted for gene replication and the transmission of genetic information.

The social basis of knowledge, therefore, does not exclude private knowledge or new knowledge. Thought may be socially conditioned, but it springs from individuals, such as Galileo or Van Gogh, who are related to the past through the present. Originally published in 1919, T. S. Eliot's essay, "Tradition and the Individual Talent," insists that no artist "has his complete meaning alone" (1962, 1502). The artist's meaning is in relation to tradition or the past. The artist must have a sense of the pastness of the past, as well as its presence. The difference between the present and the past is that the present can be conscious of the past in a way that the past could not. We know so much more than they did in the past, but, Eliot adds, "they are that which we know" (1962, 1503). While Eliot was speaking about the artist, his remarks are equally true of knowledge in general. Even the physical scientist is probing the past — the principles of evolution, the structures of stars and cells and atoms, the neural connections of the human brain, all the matter from the past, whether newly synthesized or newly split.

If you accept that knowledge is irremediably social, than there are implications for the development of expert systems. The computer is isolated. It is not in a communal relationship with human persons or with other computers, despite vast computer networks. A computer cannot know as a human person knows. Human persons come to know through a socialization process. The most important thing that we learn, namely, language, is not taught to us in the form of explicit rules. We learn language by being part of a community of language users. Collins refers to Artificial Intelligence as a "social prosthesis" because a truly intelligent computer would need to simulate a human person within a social group not simply an individual human brain (1990, 14).

TACIT KNOWLEDGE

Moreover, the relationship between what we know and what we can say about what we know is a complex and changing one. What we can say about what we know increases as more research is completed, more textbooks are written, more systems are devised for abstracting, storing, and retrieving the results of research — as knowledge is externalized. What we can say decreases as knowledge is no longer examined but enters into the systems and skills that underlie our culture — as knowledge is internalized.

The term *tacit knowledge* is sometimes used today in discussions of expertise. Polanyi developed his concept of tacit knowledge in the 1960s "by starting with the fact that we can know more than we can tell" (1966, 4). He also regarded the process of attempting to formalize all knowledge to the exclusion of tacit knowledge as "self-defeating" (1966, 20). In comprehending a skill, we must recognize a person using his body and his mind. The person is always an interior. The apprentice, according to Polanyi, learns the skill first by an "indwelling" and then by an "interiorization" (1966, 30). He or she tries to get "inside" the mind and body of the master and then interiorize the experience or bring it within his or her own interior. This whole process, of course, can only be an analogous one since a human person can never be completely penetrated or externalized.

Tacit knowledge is not unconscious knowledge. Rather it is the knowledge of a person who is able to execute a skill, but unable to say *in words* how he or she did it. John Anderson, a cognitive psychologist who has written on the development of expertise along with others such as Paul Fitts and Michael Posner, distinguishes three stages in learning a skill: the cognitive stage in which the person learns certain facts related to the skill; the associative stage in which the method for performing the skill is worked out; and the autonomous stage in which the skill becomes automatic. At this last stage, according to Anderson, "the ability to verbalize knowledge of the skill can be lost altogether" (1985, 235).

Collins takes tacit knowledge to include a wide range of cultural skills, including not only expertise, but the ability to make inductions. As we agree on how to digitize the world, "it is our common culture," according to Collins, "that makes it possible to come to these agreements, and it is our means of making these agreements that comprises our culture" (1990, 109).

Expert systems encapsulate what experts say about their knowledge. But what they say is never all that can be said. Some of what is unsaid is part of that tacit knowledge that those within the community are assumed to know. Who actually comprises that community, however, may vary. For the expert who helped to design MYCIN, for example, it might be the community of physicians or the community of microbiologists or the larger community of scientists. Some knowledge is left unsaid because it is not attended to by even the expert himself. That is why Collins insists that "the whole of an expertise cannot be captured in what can be said about it" (1990, 216–17).

Neither Collins nor Leith applies what he is saying to Third World development. That is not their area of interest. But the relevance is clear. Expertise is embedded in a community and can never be totally extracted from or become a replacement for that community.

TRANSLATION AND INTERPRETATION

The issue of exporting expert systems to the Third World is, in one sense, separate from that of designing expert systems within the Third World. In the former case, there is the added question of whether this exported expertise, which is realized and expressed in the words and concepts of a specific language and culture, can be effectively translated into the language and context of another culture.

Translation and interpretation are, in fact, central to the design and implementation of expert systems in any environment. This is clear from even a sketchy scenario of how the design process unfolds. Knowledge is represented in some way in the human brain of the expert. This internal knowledge must be circumscribed and articulated for the person designing the expert system, even if the expert and the designer are one and the same person. The system designer must interpret what is said (and unsaid), translate this into his or her own internal knowledge, and then retranslate it into the symbolic representations and knowledge structures that are compatible with specific computer hardware and software. The person using the expert system must also interpret the system's request for input, translate his response into the format expected by the system, and interpret any system output.

The impact of these processes of interpretation and translation is intensified, however, when expert systems, designed in and for one cultural context, are transported to a different cultural context. Differences in language, culture, and context mean that the designers of the technology and the proposed users of the technology may have different beliefs about the world and about themselves. A critical part of the design and implementation process for expert systems is determining what those respective beliefs are. Ultimately, expert systems represent belief as well as knowledge.

INEXPLICIT INFORMATION

Even if indigenous knowledge is explicitly represented in the knowledge base of an expert system, there is still the problem of what Robert Cummins has called inexplicit information. Cummins defines inexplicit information as information that exists in the system as the result of executing the program and not as the result of being represented in a data base or being constructed by the system (1986, 116). "A system can have much information that is not represented by the system at all," Cummins proposes, "and that doesn't function anything like knowledge or belief, even tacit knowledge or belief" (1986, 125). He makes the distinction between a rule that is represented in memory and a rule or instruction in the program that the system executes. A system with a rule represented in memory must infer what to do and must have the capacity to analyze conditions, actions, and goals. A system executing an instruction does not need to make an inference. "There is all the difference in the world," Cummins argues, "between writing a program that has access to a rule codified in one of its data structures and a program that contains that rule as an instruction" (1986, 123).

By way of illustration, perhaps you are a wildlife manager at a park with a new policy for reintroducing endangered species into the wild and maximizing their chances of survival. Coincidentally, you are working with a knowledge engineer on an expert system for wildlife management. This system will have many rules that specifically check on the size of various animal populations, the condition of the vegetation, and the predatory relationship of the animals. Based on some of these factors, an inference may be drawn that certain animals must be culled to protect the ecology of the park. One rule represented explicitly in our hypothetical expert system, might be (in pseudocode): if the population of lions is greater than x and the population of zebras is less than y, then compute lion-cull = $.05(x)$. The system would need to compare the population of lions to the value x, compare the population of zebras to the value y, and, if the postulated condition is met, compute the number of lions to cull. This same hypothetical expert system might not contain any rule for checking the population of the endangered species because, as the park manager, you "naturally" assume they will never be culled, no matter what the state of the vegetation or the population of the other animals — simply because they are already near extinction.

As a consequence, information that is not explicitly represented in the system may, nevertheless, be represented by the programmer who specifies which instructions to include in the program and the order in which the program should execute the instructions. "When we write a program," Cummins concludes, "we are theorizing about our human subjects, representing to each other the rules we suppose they execute. When we do this, what we are doing to a great extent is specifying the inexplicit information that drives human as well as computer cognitive processes" (1986, 125).

To expand on this argument, it is necessary to know how an expert system is constructed. Expert systems consist of two main components: the knowledge base and the inference engine. There is, of course, a working memory to store information elicited from the user. An expert system might also incorporate a module to reproduce its reasoning process in reaching a conclusion and make this available to the user. With the new expert system shells, the knowledge base of stored facts and rules can theoretically be tailored to the requirements of a particular environment. An individual in the Third World can use this shell and build the knowledge base to correspond to Third World "inputs."

The inference engine, on the other hand, is a given. It draws inferences or conclusions from a set of facts based on formal logical procedures, such as predicate calculus. Some of the expert systems software in use today were developed for a specific application or a specific company, using a standard language, such as C or LISP or OPS. Others were developed from expert shells. Expert shells are generally constructed by taking an existing expert system and stripping off the knowledge base. The two most frequent complaints about these expert shells are that an inference engine that works for one application does not work in a totally different application and that the method for representing knowledge in one application is often awkward or impossible in another application (Jackson, Reichgelt, & van Harmelen 1989, 2–3).

We should not be surprised that this is so. An expert in any area, whether it is medicine, metallurgy, or wildlife management, operates not only from his or her knowledge but from a community of belief. This belief consists of assumptions about the world and about oneself. A wildlife manager in the Grand Canyon National Park or in Yellowstone, for example, might believe that wildlife has value in itself apart from any commercial or practical value for humanity, that wildlife must be preserved, and that the needs of wildlife take precedence over the needs of an expanding human population or of tourists eager to capture a grizzly bear on film. The beliefs of a wildlife manager in Arizona or Wyoming will likely differ from those of one in Kenya. The latter might also believe that armed poachers should be shot and or that local herdsmen must have regulated access to parklands.

None of these beliefs will be explicitly represented in an expert system for wildlife management. That system will represent rules for culling herds and burning brush, for land allocation and introducing new species. But some of the expert's beliefs that affect decisions will be inexplicitly represented in the choice of rules that are finally selected and in the program that operates upon those rules. If the expert is not also the programmer, then the situation is even more complicated. The programmer's beliefs about the expert's beliefs about the person who will be using the expert system all come into play. Not to mention the programmer's beliefs about computers and programming.

QUINE AND THE WEB OF BELIEF

Human knowledge, upon which expert systems are based, is informed by an entire network of beliefs that can never be made exhaustively explicit — even to ourselves. Moreover, our beliefs, taken individually, are not as important to the communication process as our beliefs taken in total.

It is appropriate to speak of a community of belief because, to a large extent, our beliefs are imparted and supported by groups, such as our families, churches, professions, and cultures. Quine wrote one of his best known works about the "web of belief." His philosophy, as a whole, is based upon a behaviorist psychology. According to J. B. Watson, the chief proponent of behaviorism, states of consciousness are private and ultimately inaccessible to the psychologist. Therefore, the only scientific "data" for psychology to be based on is observable behavior. While Quine's position is clearly vulnerable to the attacks that behaviorism has sustained from the new generation of cognitive psychologists, his insights about belief, I think, are useful.

For Quine, there was an interconnectedness to our beliefs. The beliefs at the edges of Quine's web are those that are closely linked to individual sense experiences. They are expressed in what Quine called observation sentences: "Look at those dark clouds rolling in. I believe it will rain this afternoon." "The elephant over there by the baobab has a broken tusk."

"That young lion at the far end of the pride is stalking a wildebeest." Quine routinely describes observation as though it were without problems. Since Quine there has been more discussion among philosophers about the "plasticity" of perception and the need to account for variation and constancy in our perceptions of the world (Fodor, 1988; Churchland, 1988).

The beliefs at the center of Quine's web are the most theoretical. These are beliefs about physics, mathematics, logic, history, astronomy, and all the other disciplines of human knowledge. The beliefs in the center constrain the beliefs at the periphery, since our scientific heritage configures the world for us. The very interconnectedness of belief, moreover, is a partial restriction on new beliefs.

Admittedly, Quine's analogy is not entirely successful. We know that we do not consciously take inventory of all our beliefs each time we acquire a new one — sizing them up, discarding some here, fine tuning others there. And, in some individuals, their religious beliefs, for instance, can be in conflict with their scientific beliefs — even though they firmly attest both. Generally, however, a new belief must somehow fit in with the others or there must be some adjustment in our existing beliefs. This is especially true of scientific beliefs. There is some argument about whether new scientific theories really "replace" old theories. But certainly, a new scientific explanation must somehow fit into the existing scientific context. Niels Bohr, in the early days of quantum mechanics, called this the correspondence principle. In this sense at least, the revolutions of science are profoundly conservative.

Accordingly, truth, in Quine's critique, is contextual, not syntactic. For Quine, the "unit of empirical significance" is not the word or the sentence, but science as a whole. Opposed to reductionism, even in an attenuated form, Quine argued that "our statements about the external world face the tribunal of sense experience not individually but only as a corporate body" (1963, 41). The individual's observation sentence is informed by the society's body of knowledge. It is this body of knowledge that allows the individual to make sense of his or her own individual sensory experience. The individual is within the scientific context and the scientific context is within the individual. That is why, in the case of observation sentences, Quine normally expects other witnesses to the same event or situation to give their assent to the observation sentence.

Actually, Quine should have said that observation sentences are ones to which all speakers of a particular language community could give their assent or dissent if they were receiving the same stimulation as the individual making the observation. Jerome Bruner and others have suggested that situation is important in measuring cognitive performance and that "different cultural groups are likely to respond differently to any given situation" (1973, 452). Individuals from different cultural groups will vary in their interpretation of experimental stimuli and in their motivation to respond to questions. So if the witnesses to the event or situation are speakers of different languages, from different cultures, their observations may differ as well.

TECHNOLOGY EMBEDDED IN LANGUAGE

We have arrived at the center of the problem of cross-cultural communication and symbolic translation. If we really want to understand any computer technology, including expert systems, we have to start with an understanding of language. All computer technology is embedded in language. Computer software is written in artificial languages, such as Prolog, C, and LISP, each of which have different symbols, expressions, functions, and procedures, but all of which ultimately map to words and concepts in a natural language.

MYCIN, as I mentioned in an earlier chapter, is a program to assist doctors in prescribing antibiotics for patients with bacterial infections. Written in LISP, an acronym for LISt Processing, the following is a sample from MYCIN's program code (Jackson 1986, 96):

```
PREMISE: ($AND (SAME CNTXT GRAM GRAMNEG)
               (SAME CNTXT MORPH ROD)
               (SAME CONTXT AIR AEROBIC))
ACTION:  (CONCLUDE CNTXT CLASS
         ENTEROBACTERIACEAE TALLY .8)
```

This piece of code was intended to process information from the program user and return a response. If the program user entered information that the GRAM stain of the organism was gram-negative, its morphology was rod shaped, and it was aerobic, then MYCIN concluded that there was an 80 percent probability that the class of the organism was enterobacteriaceae.

Prolog, which stands for programming in logic, has entirely different constructs than LISP, although both are designed for symbolic rather than numeric computation. A classical piece of logic is accommodated in the following Prolog code:

```
fallible(x) :– man(x).
man(socrates).
?–fallible(socrates)
yes.
```

The explanation, in ordinary words, is that the two axioms — all men are fallible and Socrates is a man — are followed by the user's conjectured theorem — is Socrates fallible? The program's response is "yes, this theorem can logically be derived from these axioms."

Prolog, like all high-level programming languages, has built-in procedures that are expressed in words — assert, atom, integer, name, read, spy, write, is, repeat; abbreviations of words — nonvar, arg; or combinations of words — bagof, setof, nospy. The procedure, spy, for example, traces the execution of a specified predicate; it takes a "peek" at the system in operation. Although everything, commands and data, must ultimately be translated into the binary code of 1s and 0s, these linguistic constructs are what the programmer actually uses to code a program.

Moreover, in gathering its facts and giving its conclusions, an expert system typically employs a natural language interface in which the user must enter information at a keyboard, and select or review information displayed on a monitor or a printer. "Number of online disks?" "Is morphology rod, spherical, or spiral?" "Size of elephant population?" These might be fact-finding queries in expert systems for configuring a computer network, identifying bacteria, or managing a wildlife reserve. In some cases, this textual interface is combined with a graphic interface as well. The knowledge base that allows the system to carry on this dialogue and consultation is typically built from textbooks and from textbook-educated experts. These human experts can also supply information that the textbooks fail to mention, but this information will generally conform to a textbook model.

What is striking is that the computer's codes, knowledge structures, representations, stories, dialogues, consultations, and logic come from the context of literacy. Following this same line of sight, Jay Bolter has suggested that "the computer is best understood as a new technology of writing" — a new "species of writing" — in which writing includes not only words, but mathematics, symbolic logic, and graphics as well (1991, 9, 192). What has been offered so far as Artificial Intelligence is, in Bolter's view, basically artificial writing since the computer "is intelligent only in collaboration with human readers and writers" (193).

Computer technology, therefore, comes in many different guises, such as word processing, graphics, control systems, robotics, and expert systems. But, in the process of carrying out all these various functions, it primarily displays linguistic symbols and manipulates and transmits the binary code for these symbols. These natural language symbols and artificial language symbols must be translated in transferring expert systems between cultures in which the languages are different.

INDETERMINACY OF TRANSLATION

Over thirty years ago, Quine postulated the indeterminacy of translation (1960, ix). Since the principle has continued to be either reworked or rebutted over the intervening years, it is worth mentioning here. The principle, as Quine presented it, is best grasped through an example. Kandju is an indigenous stone-age person from the jungles of New Guinea who has had no contact with the "outside" world. He is visited, on separate occasions, by two English linguists — Farley and McDermott. In an effort to learn Kandju's language and translate it into English, Farley and McDermott observe Kandju's responses both to stimulations from the immediate environment and to their simple queries. The principle of indeterminacy of translation says that it would be possible for Farley and McDermott to produce two completely incompatible manuals for translating Kandju's language into English, that both manuals would fit all the observable "data," and that there would be no sense in asking which manual was the "right" one.

In Quine's model of radical translation, which he defines as "the translation of the language of a hitherto untouched people" (1960, 28), the utterances that are translated first are the observation sentences, the sentences that are most directly linked with sensory experience. The scene unfolds like this. A linguist is observing a woman who speaks a language unrelated to that of the linguist. A rabbit scurries past both of them. The woman says, "gavagai." The linguist has to supply sentences to the woman for her approval or disapproval. "Gavagai" might refer to the rabbit or to the whiteness of the rabbit or to a small animal or to an animal, large or small, racing across an open field, and so on. The linguist has to determine what the referent of "gavagai" is. In this particular case, it might not have been the rabbit at all.

To do this, however, the linguist must first determine how to recognize the individual's manner of communicating assent and dissent. There might be no standard expression for assent and dissent in the person's language. In some instances, assent or dissent might be expressed verbally and in other instances expressed with a gesture, a glance of the eyes, or with silence. All the linguist can do is make some guesses about the person's response from observations and then see how well the guesses work.

As I indicated earlier, Quine's whole thesis that translation must result in a systematic indeterminacy is predicated on a behavioristic approach to language in which language is simply the disposition to produce utterances in response to external stimuli. The problem for the translator is to determine what the stimuli are and to correlate these with responses. Quine does not speak to the question of how we first know whether to interpret the speaker's response as rational or irrational or how we know what constitutes rational behavior in the speaker's culture as opposed to the translator's.

John Searle has attacked Quine's basic thesis of indeterminacy as absurd because there is no empirical difference, for example, between the claim that Kandju's response meant "rabbit" and the claim that it meant "rabbit ears" or "hopping rabbit." This leaves the meaning and referent of a speaker's words to a degree indeterminate and inscrutable not only to another who does not speak the language but even to one who does — and, in fact, even to the speaker himself (1987, 130). But beyond that, Searle argues that the model of radical translation forces us to adopt a third-person point of view towards language as though this were preferable or "somehow more 'empirical'" than a first-person point of view (1987, 143, 145).

All language, Searle points out, is public. When a real-life linguist tries to translate another language into his own, he is not trying to penetrate an inaccessible object. In actual fact, he tries to figure out what is going on in the mind of the other person. He can do this because he shares with the speaker the human faculty of language.

In at least one place, Quine actually seems to be in agreement with Searle. Distinguishing between direct and indirect quotation, Quine remarks that when we quote a person directly, "we report it almost as we might a bird call," giving the details of the "physical incident" and leaving the implications to the hearer (1960, 219). Of course, in actual fact, our tone of voice and gestures might lead the hearer to imply something that the speaker did not intend, even though the words may be verbatim. Nevertheless, Quine goes on to say that "in indirect quotation we project ourselves into what, from his remarks and other indications, we imagine the speaker's state of mind to have been, and then we say what, in our language, is natural and relevant for us in the state thus feigned" (1960, 219).

The indirect quote is, therefore, a form of translation. It assumes, however, that the speaker and the person indirectly quoting the speaker share a common language. To indirectly quote someone, according to Quine, we project ourselves into the speaker and convey the speaker's "state of mind," not just his or her words. In other words, language, which is rooted in community and relationship, is the means by which we can project ourselves outward and bring the other into ourselves — and the reason we can be both speaker and hearer.

An extrapolation from this example of indirect quotation to the larger problem of communicating with a person whose language and culture are unknown to us is, I think, relevant. Faced with such a situation, Daniel Dennett would say that we must attribute rationality to the person as a first step to interpreting his or her behavior and that, through some intentional assaying, we could then begin to ascribe beliefs. Stephen Stich, Dennett's nemesis, would say that we must treat the individual as similar to ourselves. At best we can project the circumstances of this individual into ourselves and determine the state of mind that we would experience under those circumstances. The third alternative, however, would be to project ourselves into what we imagine the individual's state of mind would be under those circumstances. In reality, none of the alternatives escape a bias toward the interpreter — and that may be unavoidable.

RE-PRESENTATION

Our own human experience leads us to the conclusion that language is symbolic representation: a waterfall, a hummingbird, a wolf, a baobob, a coral reef, a snow-white spider. Something is present to us. Something exists "out there" which draws our attention by a flash of color, a buzz, a mustiness, a flutter, a warmness, a saltiness, a sting, a brush of fur against our skin. Our linguistic symbols re-present what is out there. By contrast, the linguistic and computational symbols within a computer are not re-presenting anything at all because nothing was presented to the computer to begin with. The operation of the computer upon these symbols does not depend on representation. It depends on purely formal, syntactic rules.

With our linguistic symbols, we reduce the sounds and smells and shapes and tastes to words, such as "butterfly," with attributes, such as "orange," that are part of categories, such as "insects." And we string subjects and predicates together to make statements about what is and what is not. Subjects do not exist out there. Predicates do not exist out there. Truth does not exist out there. What exists is what exists. Subjects, predicates, and truth emerge from within a human person who is radically bound to community. "That orange butterfly is a Monarch on its annual migration." We make "butterfly" the subject of an utterance. We attribute some behavior or quality to the butterfly and make that the predicate of an utterance. We determine the truth or falsity of the utterance about the butterfly. Truth is not some form or idea that is static and transcendental. Truth is a recognition process. It happens within a person.

The assumption of AI researchers to date has been that cognition involves the manipulation of symbolic representations. These representations map onto objects in the "real" world. Such a mapping was cross-cultural. The word might be "dog" or "perro" or "Hund," but the reality was the same.

But we have come to understand that words are not simple referents. Stephen Tyler, for instance, proposes that reference is a way of directing a hearer's attention when we think that the person does not understand what we are talking about. "The function of referring, then," Tyler suggests, "is less to establish an arbitrary connection between language and the world than to establish a connection between speaker-hearers" (1978, 176).

Tyler's statement, I admit, reveals a bias that most of us share. There are millions of persons who are neither speakers nor hearers, in the general definition of those words, because they are profoundly deaf. An estimated .1 percent of the world's population falls into this category (Sacks 1989, 7). While those who become deaf after acquiring language can experience phatasmal voices and sounds, the congenitally deaf inhabit a silent world. In *Seeing Voices*, Oliver Sacks explores the linguistic interaction of the deaf through sign language. Although sign language is formalized and ensnared in grammar, it is unique, according to Sacks, in its "linguistic use of space," a space that is cinematic, iconic, and mimetic, rather than simply linear, sequential, and abstract (89–121). Moreover, sign language is not a one-to-one translation of what is spoken, since the signs represent concepts and not words.

Nevertheless, the connection that Tyler describes between speaker and hearer is necessary because the meanings of words evolve. The meaning of a word is resolved in a real dialogic situation between a speaker and a listener. In our conversation, I may use a technical word you have never heard before. I may use a word that has not been spoken since the fourteenth century and is basically only familiar to Chaucerian scholars. I may use a word incorrectly. I may have three definitions in mind for a particular word and wittily intend them all simultaneously. Or I may, at that very moment, coin a word or phrase that fits the situation exactly. Some words have very precise and commonly known definitions. Some do not. We, in effect, negotiate meaning in the course of our dialogue. There have been some experiments within cognitive psychology to learn how this negotiation actually takes place (Garrod & Anderson 1987).

HERMAN WHO? HERMENEUTICS!

It is for this reason that we have seen, in the last five years, an interest of some AI researchers in hermeneutics. Even long-time AI researchers, such as Terry Winograd, have turned in this direction. For a computer to draw conclusions based on a word or a combination of words, meaning must be equivalent to a finite number of logical predicates. But Winograd aligns himself with the hermeneutists and argues that meaning derives from the immediate context of the speaker and the listener, not from a list of definitions in a dictionary. Lewis Carroll appears to have been onto this when he wrote about a girl named Alice who once fell down a rabbit-hole, only later to walk through a looking-glass — and into a world of the unexpected:

> "I don't know what you mean by 'glory,'" Alice said.
>
> Humpty Dumpty smiled contemptuously. "Of course you don't — till I tell you. I meant 'there's a nice knock-down argument for you!'"
>
> "But 'glory' doesn't mean 'a nice knock-down argument,'" Alice objected.
>
> "When I use a word," Humpty Dumpty said, in rather a scornful tone, "it means just what I choose it to mean — neither more nor less."
>
> "The question is," said Alice, "whether you can make words mean so many different things."
>
> "The question is," said Humpty Dumpty, "which is to be master — that's all." (n.d., 246–47)

For his part, Winograd relies on much loftier authorities than Alice. He concentrates on the work of Martin Heidegger and Hans-Georg Gadamer. Our conception and articulation of a real situation, according to Heidegger, is based upon the process of abstraction. In the central Western tradition which aligns intelligence with sight rather than with the other senses, Heidegger describes this process in visual imagery. When we begin analyzing our experience, we disengage ourselves and adopt a viewpoint, we focus on those objects and properties and relations for which we have terms, and, as a result, we create inevitable "blindness." The visualist Western tradition is ordinarily "blind" to the resemblance of intellectual activity to hearing, smelling, tasting, or touching. On this point, Winograd echoes Heidegger. "Reflective thought is impossible,"

Winograd concludes, "without the kind of abstraction that produces blindness" (1987, 97). The design of computer systems is equally impossible without such abstraction. The most one can do, therefore, is to be aware of the limitations.

The other philosopher that Winograd turns to, Hans-Georg Gadamer, is principally identified with hermeneutics. In his book, *Truth and Method*, Gadamer concludes that there is no method that will lead to truth, although there are methods to test what is offered as truth. Hermeneutics, so called, was originally associated with the idea of written texts, especially sacred texts. The question was this: How can I interpret the meaning of a text that was written hundreds or thousands of years ago, in a time and culture different from my own? Gadamer took the idea of interpretation beyond textual analysis and placed it at the very marrow of human cognition. "Meaning always derives from an interpretation that is rooted in a situation" (1976, 88).

WORDS THAT CONTOUR THE COSMOS

Our ability to give meaning to words, therefore, is rooted in our participation in a society, in a tradition, and in an existent situation or context. Mikhail Bakhtin, who was influenced by Gadamer, emphasizes that voices that come together in dialogue are not the voices of isolated individuals without a history or without a culture. The voices are "axiological belief systems" that can only be understood in terms of a specific social and historical situation.

All the individuals growing up in a society learn their language differently, through different speakers, and under different circumstances. I may frequently use the word "interstitial," whereas you have never heard of it. To you the word "gate" may mean "an opening in a fence." To me, it might also mean "a logical computer circuit." Words can have different denotations and connotations for different speakers in the language community.

In addition, the language is always changing because the speakers of the language are changing. Just as a dictionary is never complete, no one finishes learning a language. Expressions, such as "mall rat" and "chill out," become part of our daily conversation before they ever make it to the printed pages of a dictionary. But even though our sense experiences across our own personal histories are different, even though our manner of acquiring a language is different, the individuals in a society share, what some philosophers, including Quine, have called a conceptual scheme because of their common language. Such talk about a conceptual scheme makes some very nervous because they equate it with cultural relativism.

For the moment, it at least seems safe to say, however, that while immediate sense experience is our source of knowledge for external reality, it is language that gives coherence to this experience. It allows us to name circumscribed areas of our sense experience that we can then objectify and relativize. Something is impinging on the nerve endings of my skin, on the light receptors of my eyes, on the hairlike cells in my ears. Language helps me to define the edges, the contours, the apartness, and the togetherness of this "something" — to give those circumscribed areas of sense experience a separate reality, such as "grizzly bear" or "pinecone" or "blue iris."

Language makes our conceptualization of objects, situations, persons, and ideas possible. Sense experience is diffuse, immediate, and transitory. It is language that allows us to recall past sense experience and to think about situations that have never happened and persons or objects that have never existed. In spite of the fact that the society cannot share individual pain or joy, fear or doubt, somehow it even teaches the watchful child what these feelings are and what the socially acceptable responses to these feelings are. It gives the feelings a name.

Quine made the distinction between learning words by reference, by description, and by context. The distinction is one of degrees, not of clear-cut demarcations. Some words, such as "tiger," can be learned by pointing. Some, such as "molecule," can be learned by description. Other words, for example, "sake," can only be learned contextually — by becoming a speaker within a community of speakers. "If you won't do it for your own sake, do it for mine." "For goodness sake, don't invite him to the meeting." "Just be there for the sake of appearances." There is no way to learn the meaning of that word "sake" by reference or description. In these cases, according to Quine, it appears that we do not simply learn a word, but a part of a sentence or a whole sentence, as well as the situation in which the word was used. Clearly, we do not learn most words by learning definitions. In fact, we would be hard-pressed to give a dictionary definition of many or most of the words we use frequently and feel we "know." What we really know is precisely this: how to use them and how to understand them when they are used by others.

ENCODING SPECIFICITY

Quine's observation is supported by laboratory evidence from cognitive psychology. Endel Tulving did some experiments first with

Shirley Osler in 1968 and then with Donald Thomson in 1970 to determine how information was encoded and retrieved in human memory (1983, 211–19). These experiments led to the description of a phenomenon called "encoding specificity." Cognitive psychologists use the term *coding* or *encoding* to refer to the process by which the brain converts experiences or sensory input into whatever form is required to retain a memory of that experience. The form may be a "language," an image, a combination of both — or something quite different altogether.

In the years since his first experiments, Tulving explains that the concept of encoding specificity has become a theory "that recollection of an event, or a certain aspect of it, occurs if and only if properties of the trace of the event are sufficiently similar to the properties of the retrieval information" (1983, 223). In other words, experimental evidence suggests that we never store isolated words or pieces of information in memory. What we store are contexts.

While the earlier theories of learning emphasized the importance of association in memory recall (hot/cold, white/black, up/down), encoding specificity shifted the emphasis to context. Tulving and his colleagues presented to their subjects a list of words. In some cases, the word *cold* might be presented with a strong associate, such as *hot*; with a weak associate, such as *blow*; or without any word association at all. Tulving demonstrated that if the subjects studied the word *cold* in the context of the word *hot* or if no context at all was provided, then the subjects' recall of *cold* could be cued by the use of the word *hot*. However, when *cold* was studied in the context of *blow*, the cue word, *hot*, did not improve memory performance.

Later experiments involved the recall of words that were given context in sentences. The experiments of Tulving and Osler demonstrated that subjects' recall of words in various memory tasks improved when the targetted words were first given context by placing them in sentences. These sentences, in effect, cued recall of the words. Although these initial experiments all involved verbal recall, encoding specificity has also been used to explain memory for faces and tones. It is, of course, within our common experience that even the familiar can be unrecognizable or unretrievable out of context. Perhaps we are accustomed to seeing our conservative parish priest in black shirt, black pants, black socks, and black shoes — surrounded by Gothic arches and stained glass. If we happen to see him at a Las Vegas nightclub wearing a red silk shirt open to the fifth button, gold chains around his neck, and doing the latest steps to a Latin rhythm, we might sense that he looks familiar but still have difficulty (fortunately) coming up with his name. The dissimilarity of contexts affects our ability to recall.

THE CONTEXT OF KNOWING AND SAYING

The study of context has taken on increasing importance in the psychological literature (Davies & Thomson 1988, 1–10). There is ongoing research on context effects in memory, sensory perception, language, problem solving, and other cognitive processes. Admittedly, within this literature there is also ambiguity over the different meanings of context (Davies & Thomson 1988, 335–44). Context can mean everything, everywhere — but this is because of the always circumambient human context of language and knowing and being.

For all the foregoing reasons, knowledge representation is a central problem for expert systems and Artificial Intelligence in general. That problem is usually stated in this way: How do you structure and retrieve information in the situation where anything might be relevant? The critical point is that in the human environment of knowing, anything can be relevant. The human being, as knower, is not a self-contained system like a computer. For the human person, knowing is contextual, situational, existent, incorporative, simultaneous, experiential — a long list of adjectives that do not apply to the way in which knowledge is acquired, stored, and retrieved on the computer. The inability of the computer to use language and to coalesce and communicate knowledge in the way a human person does is clearly a severe limitation.

Likewise, the language in which computer technology is conceived, designed, implemented, and programmed creates presuppositions about its use and context. What is presented to the computer are symbolic representations of what individuals with experiential prejudices have defined as an explicit problem about which some decision is to be made concerning its truth or falsity. All those are dangerously charged words because they speak not about an "objective reality" but about a cultural mind-set that may or may not be assimilable in a different cultural environment. Not all cultures, for example, will approach what is "out there" as a problem or seek decisions about it.

SILENCE THAT SPEAKS

Silence, moreover, is as important to human communication as sound. This is true on at least two levels. The silence breaks the sound into meaningful units. If you have ever attempted to learn a foreign language, you will understand the impact of silence. When you learn the foreign language in a traditional classroom setting, you begin by learning individual words and then hearing those words slowly gathered into sentences. The first time you hear a native speaker talking in normal conversation, you often cannot recognize even the simplest words because they are not set off by silences any longer — the sounds of one word glide and blend into the next. With time you are able to catch the meaning and can sometimes anticipate the sound or reconstruct the sound when the sound itself was not heard or not attended to.

On another level, silence creates the context of what is said. The silence permits reflection on the intention of the speaker and the reaction of the listener. It adds subtleties and nuances to what is said. We have a definition for "noise" in the computer. Noise is an unwanted disturbance in the transmission of the data. It is something that must be removed in order for the data to be recognizable. Within the computer, logic values are represented electronically by two different voltage levels. When noise exceeds a certain "noise margin" that can be added to or subtracted from the logic signal, then a threshold is passed and the intended value of the signal is indeterminate. Engineers distinguish different kinds of noise — Gaussian noise, thermal noise, white noise, random noise, impulse noise. We have many terms for noise in the computer. But we do not have a term for silence. You do not need silence to manipulate decontextualized symbols with syntactic rules.

VOICES FROM THE INTERIOR

Meaning, I think we are coming to agree, is a function of context and the computer is context-free. Meaning is not predetermined; it is a matter of interpretation within an existent situation. Meaning is entangled in an existential happening that may include words but which can never be completely articulated with words.

When I communicate with another person — even if there are linguistic and cultural barriers to that communication — there is a sense in which the "other" is already in "me." Martin Buber and other philosophers have explored this I-Thou relationship. A person cannot enter into direct communication with another person's consciousness. Interpersonal communication, between an "I" and a "Thou" (or, in modern usage, "you") is always mediated by something external to consciousness, such as the body or the human voice. But the interior "structures" and predispositions (mostly subconscious or unconscious) for communication in general, and for linguistic communication in particular, are part of what we ascribe to human "nature." And in this sense, when you and I consciously communicate with one another, you are inside me in the way that a computer is not — even if this particular computer is one for which I have written programs and with whose code I am thoroughly familiar.

The computer is not a voice, not an interior addressing and being addressed by another interior. A computer cannot say "I" as a human person can. Consequently, the computer cannot project my thoughts and beliefs in the way in which another person who is speaking and listening to me can. The computer does not exist within a society, a tradition, or a context the way a human person does. And the society, tradition, and context cannot exist within the computer as all of these exist within the human person whose very language and thought assumes that the "other," the "you," is in me as I am in the "other," the "you" whom I address.

In the relationship between persons — between "I" and "you" — we are coming to a clearer understanding that knowledge and rationality are all realized in a human context that grows and varies through history and culture. That human context is largely created and expressed through language, but always in conjunction with the nonlinguistic. It is the contextual nature of knowing and reasoning that makes it difficult to successfully transport expert programs from Western cultures to Third World environments.

Chapter 7

The Limits of Logic: Folk Knowledge and Fuzz

Logic doesn't apply to the real world.

— Marvin Minsky

Logic, n. The art of thinking and reasoning in strict accordance with the limitations and incapacities of the human misunderstanding.

— Ambrose Bierce, *The Devil's Dictionary*

The last function of reason is to recognize that there are an infinity of things which surpass it.

— Blaise Pascal

It may be that the real world is just hopelessly fuzzy. Those pictures of earth from the lonely distances of space are certainly spectacular — but they are disturbing as well. To the space ship cameras, earth is all whirling clouds and swirling colors, a thistledown floating in an immense cosmic darkness. There is no clear definition of continents and land masses like we are used to on our *National Geographic* maps and globes. And even when we slip beneath the clouds for a much closer look, it is impossible to find the state lines and city limits for all the urban and rural sprawl.

Many of us pretend that our world is one of clear definitions and exact measurements. We gather confidence from the fact that our bath soap is 99.44 percent pure and that our cheese is 97 percent fat free. We know the SPF of our sunscreen, the pH of our shampoo, the mpg of our automobiles, the mHz of our personal computers, and every active ingredient in our cough syrup and our cornflakes. Even when the numbers do not all add up and the labels get confusing, they somehow make us feel better — as though we are on firm ground instead of on a spinning drop of chemistry in a universe stretching toward infinity.

But the fuzziness is there all right. We live out our lives with floating point numbers and 5/8-inch socket wrenches, with graduated mortgages and government estimates on unemployment, cancer rates, and war casualties. Some of us have even learned to be comfortable with general relativity, the uncertainty principle, and chaos.

Unable to really pinpoint and fully explain the beginnings of our universe, our species, or those abilities and accomplishments that we claim as uniquely human (such as consciousness, reason, language, writing, and technology), we have created mythic and scientific stories to fill in the gaps. None have really answered our anxious questions about life and death and meaning.

Still, most of us believe that, in the long run, everything will be okay. How long is that run? It is sort of hard to say. But that is what fuzziness is all about. There is more fuzz in our daily discourse than in all the pocket corners of every man, woman, and child in the United States.

In fact, some psychologists and philosophers, in an attempt to imbue their disciplines with the rigor of physics, have told us that what we think, and what we feel, and what we desire can pretty well be dismissed as just examples of folk knowledge. Folk knowledge is the common-sense understandings, attitudes, and beliefs that take us through our daily lives. One problem is, how common is common sense? And what exactly is common within one culture and across cultures? Clifford Geertz argued back in the mid-1960s that "common sense was a cultural system; a loosely connected body of belief and judgment, rather than just what anybody properly put together cannot help but think" (1983, 10). And our own Western common sense tells us that folk knowledge is very fuzzy where the knowledge of science is hard. The hard facts. The hard truth. One of the tasks of cognitive science is to restore validity to folk knowledge as a proper study for epistemology, psychology, and anthropology.

FUZZY EXPERTS

Where do computerized expert systems, which are formal, digitized representations of knowledge, fit into this blurred picture of the real world? And how can they be useful in Third World environments where accurate data, clear definition, and logical deduction are often not thought important — or simply not thought about at all. In many human communities, there is no equivalent of the scientific method. Within the Third World, indigenous science and technology are often a subset of folk knowledge.

Granted, there are now fuzzy expert systems — systems that incorporate fuzzy sets and fuzzy logic (Kandel 1991, 18). Some of these expert systems can even be implemented on fuzzy hardware — fuzzy computers and fuzzy logic chips. The two research programs of expert systems and fuzzy systems initially developed independently. Knowledge engineering concentrated on symbolic manipulation and plausible reasoning, seeking to symbolically represent problems and to develop ways to draw conclusions from facts that seemed to be correct. On the other hand, fuzzy systems concentrated on semantic or contextual manipulation (ways of dealing with the imprecision of language) and on approximate reasoning (ways of modeling the imprecise modes of reasoning that human persons actually use). "In all expert systems based on symbolic manipulation and plausible reasoning," Constantin Negoita explains, "uncertainty is supposed to reside in the state of our knowledge. In expert systems based on semantic manipulation and approximate reasoning, the emphasis is on fuzziness viewed as an intrinsic property of natural language" (1985, viii).

In 1965, Lofti Zadeh introduced the concept of a fuzzy set to model vague concepts or categories, such as "tall." In the Western culture, a man whose height is 6'5" is certainly tall. But where is the cut-off point? Is the man 5'10" tall? What about the man who is 5'10 1/2"? The man who is 6'5" is, in a sense, more perceptibly and decidedly in this set "tall" than the man who is 5'11", and the man 5'10 1/2" slightly more in the set than the man 5'10". Membership in a fuzzy set is a matter of degrees.

In classical-set theory, an item either belongs to a set or it does not. Such theory is founded on a logic that only permits a proposition one of two values: true or false. The Greek alphabet is a set in the classical sense, as is the U.S. Senate or the St. Louis Cardinals. You are either a U.S. senator or you are not; you are either a member of the St. Louis Cardinals baseball organization or you are not. In fuzzy-set theory, though, a proposition can be true, false, or partly true on a sliding scale. Zadeh points out that "rigor" does not play as important a role in fuzzy systems as it does in classical logic systems. "In a nutshell," Zadeh writes, "in fuzzy logic everything, including truth, is a matter of degree" (1988, 84). If the numeric value 1.00 represents "true," then it is possible to understand a statement, for instance, as absolutely true (1.00), very true (.90), practically true (.75), half true (.50), sort of true (.25), scarcely true (.05) — on and on to as many degrees as the context requires.

One very important area of fuzzy logic is "dispositional logic." Dispositions are propositions that are usually but not always true, for example, flowers need rain. Greenhouse flowers do not need rain, nor do flowers need acid rain. "The importance of dispositional logic," Zadeh points out, "stems from the fact that most of what is usually referred to as common sense knowledge may be viewed as a collection of dispositions. Thus, the main concern of dispositional logic lies in the development of rules of inference from common sense knowledge" (1988, 84).

FUSSY LOGIC

While fuzzy logic recognizes and accommodates an environment that is uncertain and imprecise, there is nothing fuzzy about fuzzy logic itself. In a sense, "fuzzy" logic is very "fussy," since it is only *programmatically* indeterminate. It is still concerned, like classical logic, with formal principles — and that means symbols, operators, connectives, and rules, fuzzy though those rules may be. "The fuzziness of such rules is," Zadeh relates, "a consequence of the fact that a rule is a summary, and summaries, in general, are fuzzy" (1988, 92).

Fuzzy or not, the purpose of an expert system is to draw conclusions from a set of facts that are coded into its knowledge base and that can be altered. The expert system must be predictive, therefore, not only for known facts or events but for new, changing, unforeseen facts or events as well. At present, expert systems attempt to do this through certain formalisms both for representing the knowledge and for solving the problem.

EXPERT FORMALISMS

As I explained at an earlier point, those who design expert systems generally make a distinction between the knowledge and the process of reasoning about the knowledge. The domain (or object knowledge) is the very limited area in which the problem must be solved, such as the selection of the best antibiotic to kill or control a determinate set of bacteria. The control (or meta-knowledge) is the strategy to actually solve the problem, generally through heuristic methods. Forward chaining, for example, is a heuristic which starts with facts and moves toward a conclusion through the repeated application of the *modus ponens* rule of inference. The Latin word *ponere* means to affirm. *Modus ponens* is a method of affirming in propositional logic according to the rule "if *p* then *q*, *p*, therefore *q*." Usually, the part of the expert program which represents the domain (the knowledge base) is separate from the part that controls the reasoning process (the inference engine).

The three most popular formalisms with expert-system designers for representing knowledge have been production rules, structured objects, and predicate logic (Jackson 1986, 30). Often, these formalisms are implemented in pattern-matching inference systems. Incoming data is matched against stored patterns, and an action is taken when the incoming pattern matches a "trigger" pattern. Production rules, for instance, are commonly called if-then rules. If a condition is met, then an action is taken. The incoming data, in this case, might be a user response to a question from the expert system. Depending on the response, the expert system might ask the user another question or display some information.

In the last few years, there has also been renewed interest in using logic as the main formalism for expert systems. It is proposed that logical languages provide the representational scheme for the knowledge and that logical deduction be the paradigm for the inference engine (Jackson, Reichgelt, & Harmelen 1989, 4). In fact, it is more accurate today to refer to logics than to logic. Within the research program of Artificial Intelligence, many alternative formalisms have been proposed because computer scientists are faced not simply with representing mathematical algorithms or propositional clauses. With expert systems, they must be able to represent uncertainty, belief, time, change, intention, intuition — a whole human context that limits, conditions, and facilitates how human persons know and judge. This has led to further experiments with multi-valued logics, fuzzy logic, intuitionistic logic, modal logic, temporal logic, and other formalisms for expression and inference which rival or extend classical logic (Jackson et al. 1989, 5; Turner 1984, 11). Like classical propositional logic, however, these newer logics are still an attempt to further formalize what is humanly experienced — since this "what" is at least partially formalized already through language.

ATTACKING RATIONALISM

All this information on expert systems should come with the caution that many of these techniques are not clearly understood, even by AI researchers, and their effectiveness has certainly not been proven. This type of textbook discussion can make it sound as though these formalisms are clearly distinct and that there are well-thought-out reasons why an expert system designer would select one over another — which is not the case. Peter Jackson, an AI researcher and author of several texts on expert systems, points out that "there is no widespread agreement in the literature as to the strengths and weaknesses of the various formalisms" nor "is there very much agreement as to which formalisms are best for which kinds of problems" (1986, 88). Jackson must finally leave the textbook language aside and acknowledge the uncertainty, the fuzziness of it all. He concludes that "knowledge is a mysterious kind of entity, about which we know remarkably little" (1986, 204). In fact, the evidence from psychology seems to suggest that human persons use different ways of encoding information and inferencing, some of which may not be "strictly logical." "We want expert systems," Jackson agrees, "to exhibit the kind of rationality we associate with human beings . . . and it may mean that future systems will be less, not more, 'logical' than they are now" (1986, 205).

Jackson is not alone in his judgment. Terry Winograd, who has worked in computer science and AI research for over twenty years, has taken an even stronger position, namely, that we need "to challenge the rationalistic tradition" itself since it has focused narrowly on only certain aspects of rationality, such as the formalization of language and knowledge, and has failed to understand language, cognition, and decision in the real world of social interaction (Winograd & Flores 1987, 8).

ETHNOSCIENCE

All of this has direct relevance for any discussion of expert systems in Third World environments. Expert systems encompass categorization and inference. There is evidence from cognitive psychology and cognitive anthropology that these may be part of a cultural code and, as cognitive processes, may be variable across cultures.

In 1973, Clifford Geertz proposed a new definition of culture as "an historically transmitted pattern of meanings embodied in symbols, a system of inherited conceptions expressed in symbolic forms by means of which men communicate, perpetuate, and develop their knowledge about and attitudes toward life" (1973, 89). While all cognitive anthropologists are not comfortable with this definition and there are ongoing discussions about terms, such as meaning, symbol, and conception, Geertz's formulation still signals a new direction in anthropology. Geertz styled his work as cultural hermeneutics, "an attempt somehow to understand how it is we understand understandings not our own" (1983, 5). Following Geertz's lead, the definition of culture, according to Roy

D'Andrade, went from something "out there" that people lived in, to something that was all in people's heads (Shweder & LeVine 1984, 7). The emphasis shifted from patterns of behavior and customs to cognition and knowledge systems.

It also created new problems for the anthropologists to solve. How do you interpret the apparently false knowledge of another culture, a culture that is technologically undeveloped by Western standards? Richard Shweder uses the example of a Bongo, from the eastern Sudan, who tells you that the ashes from the burnt skull of a red bush monkey will cure epilepsy (Shweder & LeVine 1984, 9). How do you know that this is what the Bongo really believes? Do you treat it as a false belief? Do you conclude that the belief leads to irrational behavior? Or do you, Shweder suggests as an alternative, "try to contextualize the natives' ideas or give it a frame of reference that makes it appear rational from the point of view of the persons whose belief it is" (Shweder & LeVine 1984, 9)?

Moreover, if shared beliefs can be false and institutionalized practices can be irrational, what does this tell us about the mind of the human person and the fact that, according to the enlightenment view, "all peoples are intendedly rational and scientific" (Shweder & LeVine 1984, 30)? The assumption of the enlightenment approach, Shweder points out, is "that our sensory apparatus, sensory inputs, and modes of intellectual operation are (a) already sufficiently calibrated or (b) over time can become sufficiently calibrated so as to produce convergence or agreement about what the world is really like or really ought to be like" (Shweder & LeVine 1984, 31).

Anthropologists now recognize, of course, that experience, and therefore knowledge, will vary by local ecology, culture, age, gender, social roles, and neurophysiology. Ethnoscience attempts to understand indigenous systems for defining, categorizing, and deciding, to understand the structure of knowledge and knowledge transmission within a culture.

ETHNO-LOGIC

With the technology of expert systems, we are drawn into this debate about human rationality because an expert system must somehow both represent knowledge and simulate the reasoning process of a human expert. Expert-systems design has, until now, assumed this was as we in the Western tradition perceive and define an expert and as we perceive and define reason.

Arguments about rationality are not new to anthropology (Shweder & LeVine 1984, 27–28). From the start, there have been two sides to the dispute: the enlightenment and the romantic. The enlightenment view maintains that the human mind is rational and scientific, that the dictates of reason are binding for all times and all cultures, that history is progress, and that the "primitive" is a deficient scientist. The romantic view holds, variously, that the human mind is rational, irrational, and nonrational, that it has faulty inference skills, that culture is an arbitrary code, and that the "primitive" is alogical, not illogical.

Both views have incurred accusations of ethnocentrism and cultural relativism. James F. Hamill, in *Ethno-Logic*, tries to steer the middle course. Interestingly, Hamill indicates that his goal is not directly to assist in AI research, but that he will be pleased if his work is of use (1990, 9). At the same time, he is skeptical about the current approach to expert systems which attempts to encapsulate human knowledge and skill in a set of rules, such as medical diagnosis systems, because it cannot create new knowledge or new reasoning patterns (1990, 9).

Hamill admits to a "rationalist bias" in his work since he believes that, as part of human nature, all people come equipped at birth with the ability to acquire cultural knowledge (1990, 11). He defines *ethno-logic*, a term he coined, as the study of "actual human reasoning in its linguistic, social, and cultural context" (1990, 25). Logic in modern Western culture is a specific form of reasoning that is constrained by rules of combination and consistency. It is abstract and is not directly linked to any human activity. No matter how impressive its techniques or implementations, it is, according to Hamill, culturally biased.

In an effort to avoid the ethnocentrism of earlier theories of rationality, Hamill insists that when researchers compare a non-Western logic to Western logic, they are comparing one folk system to another, not one folk system to a norm. "In some ways," Hamill writes, "the whole idea of ethno-logic is ethnocentric: it applies logic, an artifact of western European culture, to other cultures" (1990, 15).

VYGOTSKY AND THE CLASH OF SYMBOLS

Recent ethno-logic is based on the earlier work of Soviet psychologist, Lev Semenovich Vygotsky (1896–1934) and his student, Aleksandr Romanovich Luria (1902–1977). One of Vygotsky's central contributions was the concept of "mediation." It is based on the axiom that understanding requires mediation between the knower and the known, as well as between knower and knower. Signs are the forms of mediation. A sign is a meaningful symbol that has evolved within the history of a culture. The word, therefore, is a sign.

In effect, the symbols of our language mediate our perception. Perception depends on these symbolic codes to process incoming information and assign this information to categories. If the symbolic codes are different, the perception and categorization may be different as well. Codes change what they encode.

The spoken word symbolizes the reality in which the human person is immersed and of which the human person is a part. The spoken word is not identical to the reality or to the perception of the reality, since each word is itself an abstraction, a generalization. Therefore, the "world" is the given, the perceived, and the symbolized.

The written word, in turn, symbolizes the spoken word. But what is the relationship between the spoken and the written word? Is language that is written different from language that is spoken? Does the material in which the symbol is instantiated change the perception of what is symbolized? Does it matter if perception and thought are symbolized in spoken utterance, in script, in print, or in electronic dot matrices?

THE TECHNOLOGY OF WRITING

Vygotsky concluded that all major transitions or revolutions in cognitive development are associated with the appearance of a new form of mediation, a new symbol system. Writing was such a system. Paleontologists and molecular geneticists do not always agree on when modern humans evolved and whether this evolution began in one place, Africa, and spread elsewhere or occurred in many places synchronously. DNA analysis and at least some fossil evidence, however, currently point to the conclusion that modern humans — people like you and me — came upon the evolutionary scene perhaps some 100–200,000 years ago (Stringer 1990, 101; Wilson & Cann 1992, 72). At some point in that evolutionary journey, language emerged. Precisely when and how, we will never know. We do know that Mesopotamian cuneiform, the oldest script discovered to date, is less than 6,000 years old. The Semitic alphabet is less than 4,000 years old. Writing, therefore, is a late development in human evolution.

The human species may have a genetic predisposition for language, but it does not have an innate predisposition for writing. Writing, as Walter

Ong has pointed out, is a technology. To write requires tools and equipment. Of the thousands of spoken languages that have existed in the course of human evolution, only a relatively tiny number have ever been written down, and still fewer have a literature (Ong 1982, 78, 106).

For Vygotsky, there is an important distinction between the written language and the spoken language. It is not by chance that the child experiences difficulty in learning to write, even though he has already mastered the structure of his language. The difficulty is not in the motor skills or the coordination that writing demands. It is because the child has to disengage himself from the spoken word. He has to replace the spoken word with markings on a piece of paper. He has to memorize those markings, dissect them, and arrange them in a certain order on the page — an order for the letters of the word and an order for the words of the sentence. This requires a new level of abstraction.

Moreover, when he writes, he has to "speak" to someone who is absent or to an imaginary person. If there is no one there, why speak? Writing requires that he create the audience, the other, the listener. It also requires that he create the situation — the reason he is addressing these words to someone. "Written speech," Vygotsky notes, "must explain the situation fully in order to be intelligible" (1962, 100). When I speak to you, I generally do not have to explain the reason I am speaking. The context of our meeting tells you that. Perhaps we have just met and I am simply trying to make conversation. Or perhaps I approach you with a map in my hand and an expression of uncertainty upon my face. And if you do not understand the reason I am speaking or the meaning of my words, you can ask me: Is this what you mean? Is this why you're upset? Is this what you want to know? When you read a written text, the "I" of the author is not there to question or elaborate.

When the child has achieved all this, something happens. In Vygotsky's theory, the internalization of a writing system alters the structure of memory, the process of classification, the process of problem solving — all of the individual's intellectual operations. It does this because the intellectual processes have been reorganized to include this new symbol system.

LURIA ON LOGIC

One of Vygotsky's implicit assumptions was that, with literacy, human persons acquire the ability to decontextualize the use of signs. A. R. Luria attempted to test some of his theories. For his book, *Cognitive Development*, Luria did extensive fieldwork in the 1930s with illiterate and semiliterate individuals in the Soviet Union, including experiments on syllogistic and inferential reasoning. In his experiments, he did not use standard "tests." Instead he used "clinical conversations" in a relaxed atmosphere. From his experiments, Luria concluded that his illiterate subjects did not seem to operate with formal deductive procedures at all. One experiment, for example, went something like this:

In the Far North where there is snow, all bears are white. Novaya Zembla is in the Far North. What color are bears there? (Luria 1976, 107)

Typically, his subjects would reply:

There are different kinds of bears; if one was born red, he will stay that way. (1976, 107)

or

The world is large, I don't know what kinds of bears there are. (1976, 107)

or

I don't know; I've seen a black bear, I've never seen any others. . . . Each locality has its own animals: if it's white, they will be white; if it's yellow, they will be yellow. (1976, 109)

In other words, you do not reason about the color of bears. When you run across one, you take a good look. If you never run across one, why reason about one? Of what relevance are they to your practical life?

Luria also maintained that the illiterate subjects resisted all the questioner's attempts to elicit definitions. The subjects insisted that it was a waste of time to try to explain the obvious. They were asked, for example, to explain what a tree is. A typical reply was: "Why should I? Everyone knows what a tree is; they don't need me telling them" (1976, 86). When pressed to try and explain anyway, the respondent said: "There are trees here everywhere; you won't find a place that doesn't have trees. So what's the point of my explaining?" (1976, 86).

What Luria's respondents seem to be saying is that you do not need to define a tree. You can see a tree. You can point to a tree. You can run your hand over its bark and press its leaves between your fingers. You experience a tree. What has a tree to do with explanations?

Luria's research also focused on the process of categorical abstraction. He found that his illiterate subjects did not classify objects into the expected "logical" categories but into functional or situational groupings.

When asked what was different about the items hammer-saw-log-hatchet, one reply was: "They all fit here! The saw has to saw the log, the hammer has to hammer it, and the hatchet has to chop it. And if you want to chop the log up really good, you need the hammer. You can't take any of these away. There isn't any you don't need!" (1976, 58).

When pressed further with the question "What word could you use for these things (hammer, saw, hatchet)," the reply was "The words people use: saw, hammer, hatchet. You can't use one word for them all" (1976, 59). There was no attempt, according to Luria, to group the words saw, hammer, and hatchet under the category "tool." Or again, when shown a plate, a glass, a pair of eyeglasses, and a metal saucepan, the subjects did not group the plate, the glass, and the saucepan as kitchen utensils and exclude the eyeglasses. Nor did they group the plate, the glass, and the eyeglasses as "glass objects" and exclude the saucepan. Everything was included. You need the saucepan to cook your meal; you need the plate and the glass to eat your meal; and you need your eyeglasses to see your meal.

SAME OLD STORY FROM AI: "I'VE BEEN FRAMED!"

This "data" could be open to another interpretation. The grouping by situation is also a type of categorization. It is very much like Marvin Minsky's concept of a frame. The concept of a frame was developed in response to the problem of how to represent knowledge within the computer. Minsky's definition is as follows: "A frame is a data-structure for representing a stereotyped situation, like being in a certain kind of living room, or going to a child's birthday party" (1987, 96). It is a framework to which details can be added.

The frame becomes necessary because the computer, programmed with all the logical rules of syllogistic deduction, is not flexible enough to deal with real-world, meaningful situations. Certain behavior may be intelligent in one context and irrational in another. A frame is a structure for enclosing and excluding. Part of the real-world situation has be be framed out, the computer "blinded" to it. The frame is also a type of reasoning by pattern or analogy as opposed to reasoning by analysis and logic. And, of course, Minsky's frames and Roger Schank's scripts are implemented with verbal descriptions. They are not able to present the computer with images that themselves create the situation. For human persons, both images and words are able to evoke a situation or even tell a story. Roger Schank, in fact, has expanded his concept of scripts to a general theory of stories. In *Tell Me a Story*, he suggests that we are the stories we tell.

Our memory, intelligence, and sense of self depend upon all the stories we hear and tell — from "Once upon a time . . ." to the Passion of Jesus Christ.

CROSS-CULTURAL RESEARCH

In any event, Luria's illiterate subjects are very much rooted in the real world where speculation about the color of bears or the definition of a tree has no place. Their approach to classification is concrete and situational, not abstract. Luria was not saying that the illiterate subjects did not categorize. For Luria, the word itself is a generalization. So the ability to speak a language implies the ability to generalize, to categorize. The research on categorization in the last thirty years has, probably more than anything else, helped to define the area of cognitive psychology. There is growing evidence that all organisms need to categorize. The human brain receives an estimated 100 million messages per second. Categorization is required to reduce the perceived complexity of our environment and to facilitate the acquisition of new information about that environment.

However, Luria's illiterate subjects did not appear to concern themselves with abstract categorization, formal logic processes, definitions, and self-analysis. Luria does not set up a strict cause-and-effect relationship between literacy and these human abilities. For one thing, his other subjects who were literate had not only been to school but had also been introduced to the collective farms and to some industrialization. They were, therefore, engaged in new levels of social interaction that might influence their ability to reason.

What you have in Luria's book are not descriptions of controlled experiments with limited variables. The experimental method was developed as a technique for isolating and testing variables and for demonstrating a relationship between cause and effect. The blanket qualifier for repetition of test experiments and results is "all other things being equal." In *Comparative Studies of How People Think*, Michael Cole and Barbara Means go over all the problems of the experimental method within cognitive psychology and within cross-cultural studies. In the real world of human relationships and environmental interaction, all things are never equal. Even labels, such as "age" and "gender" and "culture," cover incalculable differences. How experiments are designed, how questions are asked, the expectations and motivations of the persons tested, their knowledge and training — there are all kinds of factors such as these that influence the results and interpretation of an experiment.

When that experiment is intended to measure cognitive performances, such as memory retention or reasoning or categorization, you are dealing with processes, as Cole and Means point out, that are not accessible to direct observation and are not measurable or definable in isolation (1981, 62). Consequently, not everyone would accept Luria's methods in attempting to test out Vygotsky's conclusions. However, his research does lend at least some support, if not proof, to the observation that perception, deductive reasoning, and categorization are not unmediated processes. They are in fact very complex social and historical processes that involve decisions informed by experience, culture, neural development, and literacy.

LITERACY AND LOGIC

Michael Cole introduced Vygotsky's work to the United States in the 1970s. He also conducted his own research in West Africa among the Kpelle and the Vai. Using the methods of experimental psychology, he, along with Sylvia Scribner, attempted to determine what relationship, if any, existed between literacy and abstract, logical reasoning. In *The Psychology of Literacy*, Cole and Scribner review the work of others, such as Jack Goody, Ian Watt, Patricia Greenfield, David Olson, and Jerome Bruner, who saw at least some experimental support for the theory that literacy provides the means for abstract, decontextualized thought and perhaps biases cultures toward the development of logical reasoning (Scribner & Cole 1981, 11). Their own results, however, were inconclusive and could not support any claim "that literacy is a necessary and sufficient condition" for logic and abstraction (Scribner & Cole 1981, 251).

Scribner and Cole do make some modest affirmation that their "studies among the Vai provide the first direct evidence that literacy makes some difference to some skills in some contexts" (Scribner & Cole 1981, 234). Their research was complicated, however, by the fact that the Vai live in a multilingual, multiscript environment. English is the official language of Liberia, Vai script is used for everyday practical tasks, and Arabic is read in the study of the Qur'an and in religious and liturgical practices. Moreover, Vai writing employs not an alphabet, but a syllabary that seems to have been created in the early nineteenth century by individuals already familiar with another writing system (DeFrancis 1989, 130). In a syllabary there is no direct mapping between the linguistic symbols and the sounds of the language. As a result, the writer cannot predict what the reader will read, since this partly depends on informed, but not necessarily correct, guessing (Ong 1982, 85, 88). All of these factors partially obscured the Vai research and made interpretation of the results even more contingent.

Some scholars in the humanities believe, nevertheless, that they have already marshalled sufficient evidence for at least a relationship between literacy and structured, analytical thought. Following on the work of Millman Parry and Albert Lord, which documented the way in which oral performances, including the oral epics of the *Iliad* and *Odyssey*, were woven together with themes, formulas, and formulaic expressions by persons who could not write, Marshall McLuhan explored the impact of print technology upon the human environment. "All media work us over completely . . . they leave no part of us untouched, unaffected, unaltered," and in this sense, McLuhan teased, they are a total human massage (McLuhan & Fiore 1967, 26). Alphabetic writing reduced sound to space, and print, by extrapolation, intensified the spatialization of sound. Printed symbols were standardized, repeatable, manipulable, reproducible; they were "c,o,n,t,i,n,u,o,u,s" and "c–o–n–n–e–c–t–e–d" (McLuhan & Fiore 1967, 44). As such, they confirmed and extended a sequential organization of thought.

Moreover, the printed text could be regarded as closed and final in a way in which the manuscript was not, since, generally speaking, manuscript copies were never identical and words were never locatable in the same "space" on the page. The printed text allowed a level of analysis that had not previously been possible with the spoken word or with the manuscript and created a fixed point of view (McLuhan 1962, 56). Also, the portability of the printed book contributed to a new cult of individualism by creating the "solitary reader" and allowing the human person to distance himself further from the social context (McLuhan 1962, 206–08). In contrast, the interdependence fostered by electronic media, such as television and computers, "recreates the world in the image of a global village" (McLuhan & Fiore 1967, 67). These electronic media, in McLuhan's lingo, force us to "live mythically" — to live in the present and participate more fully in the whole human environment through the interplay of instantaneous and simultaneous information — even though we continue to "think fragmentarily" as print technology, its texts, and our whole typographic culture have trained us to do (McLuhan & Fiore 1967, 114).

Despite McLuhan's popular style and fondness for overstatement, Eric Havelock saw his insights as "pioneering." Havelock, a classicist at Harvard, examined the differences in the "literature" from Greece during the time of Homer (c. 800 BCE) and during the time of Plato (428–347

BCE). In *The Republic*, Plato lashed out against mnemonic poetry, attacking a tribal identification with the epic heroes and appealing instead for more abstract, analytical thought. Eric Havelock pursued Plato's critique of the oral tradition, maintaining that Greek civilization underwent drastic changes as the practice of writing developed. The *Iliad* and *Odyssey* were originally oral performances that were later written down. Homer, to use Albert Lord's phrase, was a singer of tales, not an author.

The formulaic phrases and cyclical composition of the Homeric poems were a paradigm of communication in an orally based culture. They were a means of communicating culture by making it memorable. Mnemonic poetry, in other words, was a structure for information storage and management. "Between Homer and Plato," Havelock writes, "the method of storage began to alter, as the information became alphabetized, and correspondingly the eye supplanted the ear as the chief organ employed for this purpose" (1963, vii). The underlying proposition here is that the way we use our senses is intimately connected to the way we think. In the aftermath of the invention of the Greek alphabet, an act of hearing was slowly being replaced by an act of seeing as the primary means of communicating and storing information. Writing gradually made it possible to do away with the oral formulas and the epic heroes and replace these oral narratives with analytical thinking.

THE CONFIGURATION OF THE SENSES

Beginning with his scholarship on the educational reforms of Peter Ramus and the impact of those reforms on Western methodologies for visually organizing knowledge of a "spatialized universe," Walter Ong, like Havelock, has also focused on the changing role of the senses in the management of knowledge. For Ong, the "sensorium" is a particular configuration of the mind's sensory awareness (1967, 1–16). The senses can be configured in different ways. When one sense dominates, it becomes a paradigm for grasping reality. Which sense is the dominant one varies through time and culture. Changes in the sensorium are particularly evident, Ong explains, in the contrast between oral-aural cultures and print-oriented cultures (1982).

Accordingly, in an oral culture the dominant sense is hearing. Visual perception, no matter what the time or the culture, is, of course, a critical sense. Our earliest records of human consciousness are stamped in delicately fashioned stone tools, antler artifacts, and amazingly powerful drawings of bison and reindeer on the walls of Lascaux, Altamira, and other prehistoric caves of Western Europe. These all attest to early man's visual acuteness. Scholars, such as Havelock and Ong, are not denying the importance of the visual sense to man's development — even to his survival. What they are saying is that, in an oral culture, information is primarily communicated and "managed" through the sense of hearing. The meanings of words are not stored in dictionaries. The knowledge of the culture is not stored in books, although some of it may be captured in images. In the oral culture, knowledge is primarily dependent upon existent human memory. Knowledge is incorporated into formulas that will facilitate that memory, including proverbs, ritual chants, and oral tales.

TRANSFORMATIVE TECHNOLOGY

Scholars such as McLuhan, Havelock, and Ong came out of a humanistic tradition and drew their conclusions largely from the study of historical texts. When they maintain that the technologies of writing and print have transformed human consciousness, they are not referring to a clear cause-and-effect relationship. Even hard-core scientists are wary of speaking about causes and effects unless they can strictly control their "variables" under experimental conditions. And still, relationships can be suspect, as the scientist and the subject bring unconscious biases to the experiment. When you are talking about phenomena that are enfolded in the whole sweep of human history, it becomes impossible to isolate cause and effect.

Nevertheless, one of the important and still controversial observations of this particular line of scholarship is the "chirographic base of logic" (Ong 1982, 53). Accordingly, the invention of formal logic, starting with the syllogism, required, first the interiorization of alphabetic writing and later the increased formalization and abstraction afforded by print technology. Jack Goody, a social anthropologist who has also concentrated on the impact of writing upon thought, suggests that although writing was indeed a precondition for Aristotelian logic and the invention of the syllogism, the mere acquisition of writing does not produce logical thinking unless it is accompanied by a whole written tradition that embodies such thinking (1987, 226).

Yet, that written tradition and our everyday reasoning is not as "logical" as we like to think it is — and it never was. Aristotle's own name for logic was *analytiké*. It was not conceived as a separate "science," but was intended to be a tool of philosophy, and thus the name *Organon* (meaning tool or instrument) was applied to a collection of his logical treatises. Aristotle's logic was basically a logic of terms, more specifically a logic of common referential nouns.

Every A is B.
Every B is C.
Every A is C.

The Aristotelian syllogism was really an if-then propositional expression: if A and B, then C. Aristotle is generally attributed with introducing the variable to logic, so that the variables A, B, and C could stand for any noun expression. This type of rigid demonstration, or *analytiké*, was distinguished from *dialektiké* which was a method for arguing — generally in dialogue with another person — with probability, not certainty, on any given subject. Much later, especially in the Ramist tradition, dialectic tended to be equated with teaching, so that teaching something was practically the same as demonstrating its validity (Ong [1958] 1983, 156).

BIAS IN REASONING

Quite apart from this pedagogical bias that runs through the Western tradition, Jonathan Evans has detailed evidence for bias in everyday human reasoning. In addition to Tversky and Kahneman, he cites the work of Wason, Nisbett and Ross, Johnson-Laird, and others. Evans confirms that, for the most part, we "hold false beliefs about our own cognitive processes" and that "these false beliefs are self-flattering" (1989, 109). Evans defines a bias as "a systematic tendency to take account of factors irrelevant to the task at hand or to ignore relevant factors" (1989, 9). Some of our biases are the result of "availability," i.e., the information which is actually thought about when the subject is reasoning. In some cases, individuals will not have access to pertinent information or will not consider it in making their decisions. In other cases, certain "vivid" concrete information will outweigh dull, abstract information (Evans 1989, 27). There is also what Evans calls the confirmation bias, which is that human beings have a tendency to search out information that is consistent with their beliefs or hypotheses and to dismiss information that is not. And finally, the attention span and working memory of all of us are limited — a limitation that affects our reasoning process. So even we do not fit the "model."

Still, there are continued attempts to determine whether others do or do not reason as we do in the West. Between 1975 and 1976, Edwin Hutchins studied reasoning patterns among the Trobriand Islanders of Papua, New Guinea. He did this by examining the inferences and logical arguments that were used in legal disputes over land tenure. The assumption had always been that the real world of social interaction was too messy to study directly so artificial situations were created. Unlike past experiments in which logical syllogisms, for example, were presented to the subject, Hutchins's analysis "was performed on naturally occurring discourse in a society and language very different from our own" (1980, 126). Land tenure was selected as the focal point since, in the Trobriand society as in most nontechnological cultures, land assumed utmost importance as the source of wealth and power (1980, 15).

Hutchins compared Trobriand reasoning patterns to what could only be described as an ethnocentric model. "The clear difference between cultures with respect to reasoning," Hutchins construes, "is in the representation of the world which is thought about rather than in the processes employed in doing the thinking" (1980, 128). In other words, within the world of sense experience, there are an infinite number of perceptible things. The individual must be able to map this infinite universe of experience onto a finite number of culturally and linguistically defined categories. These categories of experience are what people must have in order to reason about their experience. For Hutchins, therefore, the differences were in categorization, not in reasoning (1980, 128). In drawing his conclusion, Hutchins acknowledged an intellectual debt to Eleanor Rosch and her research on categorization.

CATEGORIES AND COGNITION

Writing in the 1950s, Benjamin Lee Whorf proposed two hypotheses: first, that linguistic structure determines cognitive structure; and second, that cognitive systems are different in speakers of different languages. Whorf's hypotheses have proved very seductive — and very difficult to subject to empirical testing. In the intervening years there have been philosophical debates, psychological experiments, and anthropological field work attempting to prove and disprove them.

The study of how persons in different cultures categorize colors has been a popular area for studying Whorf's hypotheses about the influence of language on perception and categorization. Berlin, Kay, and Rosch are a few of the names prominently connected with this research. The perception of color is a universal human experience, except for those who are colorblind. Color is a physical phenomenon that is related to the wavelength of light — as well as to the neurophysiology of the eye and brain. Brent Berlin and Paul Kay challenged the hypothesis that speakers of different languages actually perceive colors differently and that the color categories are arbitrary. Their research identified eleven basic color categories: white, black, red, green, yellow, blue, brown, purple, pink, orange, and grey (Berlin & Kay 1969, 2). Some languages, including English, have all eleven, while others have as few as two.

All the informants, with one exception, resided in San Francisco, so that sometimes the researchers only had one informant for a language and many of them were bilingual university students (Berlin & Kay 1969, 7). This clearly attenuates the research results. Nevertheless, the languages of the informants were diverse: Arabic (Lebanon); Bulgarian (Bulgaria); Catalan (Spain); Cantonese (China); Mandarin (China); English (United States); Hebrew (Israel); Hungarian (Hungary); Igbo (Nigeria); Indonesian (Indonesia); Japanese (Japan); Korean (Korea); Pomo (California); Spanish (Mexico); Kiswahili (East Africa); Tagalog (Philippines); Thai (Thailand); Tzeltal (Southern Mexico); Urdu (India); and Vietnamese (Vietnam) (Berlin & Kay 1969, 7). Essentially linguistic investigations led to the conclusion that there was a particular ordering to the basic color terms. If a language encoded fewer than the eleven categories, according to Berlin and Kay, there were limitations on which categories it would encode (1969, 2). Thus, a language with only three basic color terms would include black, white, and red. If the language had four basic color terms, the possibilities were either black, white, red, and yellow, or black, white, red, and green (1969, 2). Although the boundaries between the color categories were said to vary from language to language, the focal colors, or best examples within a color category, were found not to vary significantly among the different languages. Berlin and Kay concluded that there were "semantic universals" in color terms that could "be termed evolutionary" (1969, 1).

Eleanor Heider Rosch found that the Berlin-Kay research coincided with her own work on Dani, a language in New Guinea that reportedly has only two color categories: *mili* for dark "cool" colors and *mola* for light "warm" colors. While having only two color categories, the Dani, in memory-recall experiments, generally recognized focal colors more readily than nonfocal colors. Focal colors corresponded to what Rosch in later years termed prototypes, or members of categories that were more representative of a category than other members. Interestingly, some of Rosch's conclusions echoed earlier observations by Wittgenstein that features alone did not determine categories. In any event, Rosch concluded that "basic color terminology appears to be universal and that perceptually salient focal colors appear to form natural prototypes for the development of color terms" (1987, 274). Rosch believes that her work on color perception and that of others, such as Paul Ekman on facial expressions, points to the fact that there are "natural categories" and that the Whorfian hypotheses do "not appear to be empirically true" (1987, 275, 277). Some hedging of bets is obviously still required because cross-cultural research in all areas has been very limited.

EMBODIMENT OF REASON

George Lakoff has reviewed the seminal research on categorization in a lengthy book with a title that almost sizzles: *Women, Fire, and Dangerous Things*. The title was prompted by the fact that Dyirbal, an Australian aboriginal language, has a category which, in fact, includes women, fire, and dangerous things. In the classical view, Lakoff explains, categories were like "containers" in which things with common properties were mentally placed together. Now, while some men might recognize the "logic" of putting women, fire, and dangerous things in the same category, the conjunction should normally seem strange to us. Why then would the aborigines have such a category?

Lakoff argues that categories are a window into the mind, since categorization is basic to perception, grammar, and logic. His book disputes the "classical" view that categories exist in the world independent of people and are defined only by the characteristics of their members and not by any characteristics of persons themselves. He also challenges the idea that reason is a transcendental and disembodied faculty that can be characterized by the manipulation of abstract symbols (1987, 8). Rather, for Lakoff, categorization and reason have very much to do with the human body — not only with the neurophysiological capacity to perceive, but also with the capacity to imagine and to make use of metonymies, metaphors, and images.

One of the criticisms of cognitive science has been that it only makes marginal reference to the biology and neurophysiology of the brain. In its efforts to describe the mind's principles of organization, cognitive science has paid scant attention to the fact that the mind is embodied, that the brain was part of millions of years of neural evolution, and that mind arose at a point in that evolution (Edelman 1992, 14). Lakoff recognizes that the study of categorization is the key to the study of reason, however, embodiment is one of his recurring themes. "Reason," Lakoff emphasizes, "is embodied in the sense that the very structures on which reason is based emerge from our bodily experiences" (1987, 368).

SHATTERED REASON, SHATTERED WORLD

Taken to its conclusion, Lakoff's position is that rationality is not one but many. He is aware that he has left himself open to the charge of cultural relativism by suggesting "that there may be many conceptual systems as reasonable as our own around the world" (1987, 304). Physicists and physicians, Lakoff counters, often make use of alternative conceptualization, but the idea only becomes threatening "when we talk about how people living in unfamiliar cultures and speaking unfamiliar languages think" (1987, 306).

There are many degrees of relativism even among outspoken relativists. Lakoff prefers the term "experiential realism" to characterize his own brand because it emphasizes a commitment to the existence of the real world with all its constraints and reflects the idea that thought grows out of embodiment, that experience includes not simply perception and the actual or potential experiences of individuals or communities, but "the internal genetically acquired makeup of the organism and the nature of its interactions in both its physical and its social environments" (1987, xv).

Stephen Stich has, in turn, given the philosophical arguments for cognitive pluralism in *Fragmentation of Reason*. He suggests that perhaps we acquire both our language and our inferential systems in a parallel mode. "If that were true," Stich concludes,

> Then it might well turn out that neither our entire inferential system nor any part of it is innate or universally shared among normal humans. The existence of innate inferential strategies is an entirely open, empirical question, as is the extent to which a person's inferential strategies are shaped by his or her cultural surroundings. (1990, 72)

Even if there are parts of our cognitive system which are innate, Stich argues that "there is no biological or evolutionary reason to think that those parts do not differ markedly from person to person or from culture to culture" (1990, 74). Stich's own account of reasoning is pluralistic, relativistic, and pragmatic "since it entails that different systems of reasoning may be normatively appropriate for different people" and that all "cognitive mechanisms or processes are to be viewed as tools or policies and evaluated in much the same way that we evaluate other tools or policies" (1990, 14, 24). For the pragmatist, there is no best way of reasoning or best way to improve reasoning. Stich proposes that "quite different strategies for improving cognitive processing may prove preferable for different people with different goals, different technologies, or different environments" (1990, 158).

I have detailed some of the old and new research and reflection on logic and rationality simply to show that there is clearly no agreement about what is rational, what is irrational, what is universal, and what is relative in this debate about culture and cognition. The lack of agreement, however, has clear implications for the potential role of expert systems in Third World development — and it can open our understanding to more than purely technical considerations in implementing this technology.

IMAGERY AND INFERENCE

We do not reason about the world. We reason about our representations of the world. Often we equate this reasoning with some abstract process that takes place within the human brain. This reasoning, however, may involve imagery, for instance, which we have retained or which we generate (Kosslyn & Koenig 1992, 145–55). Some of the great scientists, including Albert Einstein, describe using mind experiments to reason to their conclusions; the mathematical demonstrations came later — and sometimes from others. Obviously, the imagery that we evoke will depend on many factors, including culture, neurophysiology, gender, and so on. If deductive reasoning and inference are variable across cultures — whether because of differences in the reasoning process or in the representation of the world that is reasoned about — we must raise the question of how knowledge is represented and how inferences are actually made in the expert system.

Some of the research on the development of communication media in the Third World and the interaction of these media with long-standing social and political authority can be useful starting points as new questions are posed and answers sought (Sreberny-Mohammadi in Gronbeck, Farrell, & Soukup 1991). What happens when expert systems — an extension of the technology of literacy — are introduced into an environment of residual orality, an environment that still persists in many developing areas where illiteracy rates are high? In what sense are Western problem-solving approaches and Western inferences simply imposed upon the indigenous knowledge base of an expert system? Is such an imposition justified? Who assumes responsibility for the network of inferences that is built into the expert-systems technology transported to developing nations? These are not trivial questions since expert systems can result in decisions that will affect Third World environments and cultures.

Expert systems have built-in limitations which they share with all other AI implementations: (1) AI researchers do not even agree among themselves whether it is even possible to symbolically represent knowledge in a computer system so that it can select out the relevant information in a given context and respond in some appropriate and adequate way; (2) human intelligence is not always rational; (3) the rules by which human experts make decisions are never wholly explicit. If this all sounds fuzzy, it is. And the blur can only get worse when we attempt to implement expert systems in Third World environments.

THE ELECTRONIC ANCESTOR

As I indicated in the previous chapter, in the Third World setting, local knowledge about the environment, about agriculture, about healing, and so on, may be incorporated in proverbs or rituals. A few years ago, there was an eye-catching article in one of the popular news magazines about attempts to computerize the irrigation rites of Bali. For a thousand years, the rice farmers of Bali would appeal to Jero Gde, the high priest of the water temple, when they needed water for their crops. The Indonesian government and the Asian Development Bank decided to design a new system.

From 1978–1984 they are reported to have spent $24 million building dams and canals, and introducing different rice strains and pesticides. The project was a failure. An anthropologist from the University of Southern California, Stephen Lansing, suggested that the project's engineer talk to the waterpriests. As it turned out, the priests of the water temple were "experts" in resource management and it was their knowledge, incorporated in the temple rituals, that for centuries had successfully determined planting, irrigation, and harvesting schedules and had preserved the nutrients in the soil, while keeping rats and insects from spreading. A computer model proved it to be "one of the most stable and efficient farming systems on the planet" (Cowley 1989, 50).

Anthropologists today recognize the efforts of the AI community to develop expert systems as similar to the long-time goal of ethnographers in representing cultural knowledge (Guillet 1989, 57). For many years they have encountered and written about the difficulties in capturing and formalizing indigenous knowledge and some of them approach this new technology of expert systems as a means of preserving and disseminating this knowledge for the benefit of the Third World itself. "The ability of expert systems to formalize, or at least make explicit, tacit, often unconscious, knowledge in a production rule form," writes David Guillet, "facilitates the capture of successful indigenous knowledge and its diffusion to people of similar language and culture who have lost this knowledge" (1989, 58). Even if knowledge cannot be completely formalized, some effort, however flawed, is required to preserve indigenous knowledge and expert systems might provide the best means available.

Still, the problem will always be how to extract local knowledge and local reasoning from a context that is dense with religious, social, and cultural significance. To do this will require the local "experts" themselves, looking critically, perhaps for the first time, at their own knowledge and their own roles within their cultures. It will also require us to set aside any convictions that our own rationality is the measure of reason and humanity.

Chapter 8

The Wound of Knowledge

It hath been taught us from the primal state
That he which is was wished until he were.

— Shakespeare

Loren Eiseley submitted that this line from Shakespeare was the "deadliest message" that we will encounter in literature because it thrusts upon us "inescapable choices" and pierces "the complacency of little centuries in which, encamped as in hidden thickets, men have sought to evade self-knowledge by describing themselves as men" (1971, 55). For this paleontologist who fingered the bones and stone artifacts of our earliest ancestors, we are, each one, just a "dark, divine animal" journeying to become human.

As a result, we are inevitably fragmented and make-believe. All this must somehow be remembered when we start to talk so confidently about knowledge, knowledge systems, rationality, culture, First World and Third. Where once our human ancestors stood in a glacial field and tossed carved animal bones upon the ground to read the future, we have come with our technology to a new field of terrible freedoms. And our technology might give us the world we wish for.

The title for this chapter was taken from a book of the same name by Rowan Williams, which is, in part, about the "profound contradictoriness" of Christian faith: to be rich, you must be poor; to be free, you must surrender your will; to know, you must follow the way of unknowing; to live, you must die. Perhaps because of that Christian influence on Western culture, there has also been an ambivalence in our attitude toward knowledge. The truth will make you free, but knowledge is the end of innocence. Let us be clear that the condition under which many persons live in the Third World is not innocence, but poverty. Yet, there is a lingering sense that it is difficult to alleviate that condition without wounding the indigenous cultures with our knowledge and technology. Life is constant transformation — neither individuals nor cultures can rest in stability.

This book has been an exploratory journey and in these last few pages I simply want to turn around and spot our tracks from this new vantage point. At the outset I emphasized that I was interested in expert systems as an emerging technology with all its current limitations and experimental formalisms. My intent was not to enter the ongoing debate about whether computers and their software could be labeled "intelligent."

There are questions about the appropriateness of expert systems to the immediate human priorities of the developing nations. Taken just as they are, expert systems do not represent *the* solution for Third World problems. Those problems are diverse and complex. In some cases, very basic needs have to be met first: available and nutritious food, clean water, sanitation, hygiene, housing, fuel. The needs for freedom, justice, and knowledge are, in another sense, just as basic.

Moreover, the human realities of poverty, injustice, illiteracy, economic and political instability — all the conditions that we commonly use to differentiate the First from the Third World — will vary from Latin America to Africa to Asia. Terms such as *Third World* are ambiguous even though they give us a mistaken sense that we have snared reality in our definitions.

In addition to the great cultural and religious diversity within these developing nations, there is a marked difference in their degree of indigenous technological preparation and in their potential for participation in the development or implementation of expert systems. Still, I would agree with those who argue that computerized knowledge systems can make a contribution to development in Third World environments.

It does not seem an exaggeration to say that expert system technology is unique in relation to any technology that has preceded it. The design of earlier technologies resulted from the application of human rationality to the solution of particular problems. Expert systems, however, purport to externalize man's reasoning ability itself, to alter modes of decision, and to cultivate knowledge outside of the human mind. They are a potential threat, in a very subtle way, to personal and social identity.

In his novel, *Things Fall Apart*, Chinua Achebe describes the disintegration of Igbo society at the end of the nineteenth century through the eyes of Okonkwo, one of the village elders. The title of the novel is from a poem by William Butler Yeats, "The Second Coming," that speaks about the dissolution of civilization: "Things fall apart; the center cannot hold." With the intrusion of British colonialism and the Christian missionaries, the whole nucleus of Igbo life — traditional practices, beliefs, relationships — is split apart. Achebe does not ignore the evil in the Igbo traditional society, such as ritual killing, but he continually points up the integrity there as well and the conflict between African modes of thinking and reasoning and those of the Europeans.

Whether in a traditional society or in a modern society, our identity and role are largely determined by what we know. To convince someone that most of their knowledge is useless or even wrong is to assault that individual at the very center of his being. It is likely to cause interior collapse. In *The Man with the Shattered World*, A. R. Luria writes about a brilliant young man named Zasetsky whose life was broken into bits and pieces when a bullet penetrated his brain during a battle of World War II. As a result of the brain injury, his memory, his grasp of language, his ability to recognize objects and to make relationships were all damaged. After many years, he was able to recover some use of language and to record his uncentered thoughts in a journal. "All my knowledge is gone," Zasetsky wrote (Luria 1972, 139). This loss was "catastrophic" because it wiped out all of his past and left him with an inability to make sense of present events and perceptions. The subsequent incomprehensibility of the world and of himself induced in Zasetsky an unbearable "fatigue and loathing," a dying from within.

If expert systems are simply offered to the Third World as a substitute for their indigenous knowledge and expertise, we are, in a very real sense, offering death, not development. Technological innovation is not as important for Third World cultures as their growing consciousness of choice and the exercise of their own rationality to direct their own development.

For those who will be the teachers and disseminators of this technology, it will be important to legitimate local knowledge and elicit local community participation in the design and implementation of expert systems. This participation must not be token. The experts must be open to the possibility of their own ignorance and not let their self-perceived expertise preclude local participation and decision. In fact, there must be a recognition by the knowledge engineers that effective design will actually depend on local knowledge, local review, and assessment. It is difficult to envision how expert systems will be beneficial without this.

This legitimation of local knowledge, however, will not be an easy task. In some cases, the local knowledge may only reside in women as part of their cultural role. This may include knowledge about animal husbandry or fuel resources for cooking and heating. In other cases, the knowledge may be reserved for elders or religious healers. They may hold unique local knowledge of medicinal plants and remedies. There may also be reasons that these individuals do not want to reveal their knowledge — internal power structures or religious prohibitions.

What is required, moreover, is not simply the eliciting and extracting of local knowledge. There must be an understanding of how this knowledge is encoded, acquired, and transmitted and of how Western perceptions of knowledge and expertise differ from those of the local culture.

This local knowledge may be imprecise. It may be expressed in stories, in initiation rites, in village dances and dramas. It may be communicated in proverbs. Proverbs are a very common form of speech in African societies. "The proverb," according to one Igbo speaker from Nigeria, "gives credence to what you are saying; it is quoting the experts" (Penfield 1983, 3). Joyce Penfield has shown that these proverbs are often ambiguous outside of "a specific interactional context" (1983, 7). To understand what is being said, you must understand the context.

Researchers in both cognitive psychology and in Artificial Intelligence seem to be coming to the same conclusion. Within cognitive psychology there is a whole line of research on "context effects," and AI researchers continue to experiment with variations of frames and scripts to represent knowledge for the computer (Cohen & Seigel, 1991; Davies & Thomson 1988; Tulving 1983).

This new emphasis on context is grounded in the understanding that there is a social and a cultural dimension to knowledge that has been transparent to us. The language by which we express our knowledge is social to its very root. We acquire it from others. Also, even when we are most articulate, we never say all that can be said. Tacit knowledge is left unsaid because others in the community are assumed to know it or because it is not attended to even by the speaker. Expert systems encapsulate only what experts say about their knowledge, not the knowledge itself as it exists within unique human interiors.

All the distinct features of knowledge within the Third World context will be pitted against Western prejudices that "real" knowledge, "technical" knowledge is formalized, logical, verifiable, externalized, impersonal. As we begin to examine these prejudices, though, some evidence is pointing to the conclusion that our knowledge, the categorization upon which our knowledge is based, and our rules for reasoning about this knowledge may be variable across history and culture.

Moreover, in the West, we have often gained knowledge at the expense of wisdom. In traditional societies, wisdom must be achieved and, therefore, it is often valued over education, wealth, or physical strength. The elders are expected to possess wisdom to a greater degree than the younger members of the community. Wisdom has to do with playing one's social, interpersonal, and interactional roles successfully. It is seeded in the culture itself and is expressed in qualities that allow one to survive, to form beneficial relationships, to manage conflict, to present one's views persuasively, to consistently make sound judgments, and to advance one's position in the culture — all with exceptional skill. One can have knowledge without experience, but one cannot have wisdom without experience. "Wisdom," according to Aeschylus, "comes by suffering." In short, wisdom is all that formalized, scientific knowledge is not. Wisdom is insight into knowledge itself. It is knowing how little can be known and that all of human knowledge cannot protect you from death.

While respect for local knowledge and wisdom is critical, it is also important not to idealize or romanticize local knowledge and insist that Western technology in a non-Western culture must always be exploitative and destructive. The local knowledge is dynamic as well. In some cases, traditional solutions to environmental challenges do not hold up as those challenges intensify and change. Western knowledge may provide better solutions that are still acceptable to local culture and customs. There is an unshakeable sense that Western science and technology are moving inexorably around the earth and that other systems of knowing may contour them, but cannot avoid confronting them.

Traditional systems may be improved, therefore, by incorporating certain elements of Western knowledge and science. This selective assimilation of Western culture already occurs throughout the Third World. As a result, it may even be misleading to use the terms *indigenous knowledge* or *local knowledge*. Some of the knowledge in the local community may have been acquired, recently or in the past, from neighboring nations or from Western concepts or terminology. Wole Soyinka, the Nigerian author who won the Nobel Prize for literature in 1986, refers to those who insist that there is an untouched African culture as Neo-Tarzanists (Soyinka 1991). With global telecommunications, international travel, visiting researchers and consultants, knowledge is free-floating. Notably, Western culture — including its science — has continually assimilated elements from East and South. This includes concepts, such as the Hindu-Arabic numbering system, as well as brainpower. Some of "our" most distinguished scientists are originally from developing nations. While they are routinely listed as U.S. Nobel Prize winners, for instance, Har Gobind Khorana, a chemist who won the award for medicine in 1968 for his interpretation of the genetic code, and Subrahmanyan Chandrasekhar, an astrophysicist who won the award for physics in 1983 with his studies on the structure and evolution of the stars, were both born and initially educated in India — though their later education and research were, in fact, in British and U.S. universities.

In the preceding chapters, I referred to the two basic strategies for the introduction of expert systems into the Third World: technology transfer and local design. It would be a mistake to assume that a technology such as knowledge-based systems can be started or stopped by timing and planning. One problem with these seemingly rational approaches to technological strategy is that the development and the direction of a technology are not, in themselves, rational processes.

It is not always possible, moreover, to save a developing nation from the mistakes that were made earlier by its industrial neighbors; to protect a developing nation from the cultural values that accompany a new technology; or to completely orchestrate equity and justice and development. We can set goals, but those goals must take their chances in a real-world setting. We only have limited control of the boat — not control of the river.

Clearly, there are critical limitations to the transfer of expert systems from the Western industrialized nations to the Third World. This does not mean that the export of such technology is to be discouraged or that this technology will be completely inappropriate or ineffective in Third World environments. It does mean that some of the problems it will create are predictable and that some choices for the technology will be better than others.

The alternative to technology transfer is that the expert systems be designed within the Third World when local knowledge and technology permit this approach. If either strategy is to contribute to meaningful development, however, one thing is of paramount importance. The design of these expert systems must incorporate indigenous knowledge and indigenous reasoning about knowledge so that their "solutions" and "decisions" can be those of the local communities, especially those communities that have been marginalized from power and from privilege, and that have, consequently, maximized survival techniques.

The decision to implement knowledge-based systems places ethical demands and risks upon those in power within Third World nations and upon those from the industrialized nations that trade with and for that power. Any assessment of this new technology requires ethical and not simply economic tools of analysis because there are ethical issues in withholding, sharing, negating, and using knowledge. Many — perhaps most — may not be able to rise to these ethical demands. Knowledge is a threat to power. It has also been a routine means to withhold power from the poor.

Yet, without grassroots participation in the design and use of this knowledge-based technology, there can be no lasting benefit and no empowerment for the poor of the Third World. And if this human empowerment of the poor is not the ultimate goal of our international development efforts, then all our technological knowledge, including our attempts to recreate ourselves in silicon and symbols, diminishes us and puts wisdom and global peace further from our reach.

Certainly, tackling any topic as rich as this one leaves the writer open to the criticism "when do you know you are finished?" Paraphrasing Paul Valéry, the French poet and essayist, I suggest that books are never finished but abandoned. As in good conversation, some omission of the details prevents boredom — in one's readers. More importantly, we all work within limitations of various kinds. Some of these are the limitations of time, knowledge, and communication. It remains inescapable that no matter how much we say about a subject, there is always something more that can be said. That is one of the shared handicaps of experts — human or not.

Bibliography

Achebe, Chinua. *Things Fall Apart*. New York: Astor-Honor, 1959.

Adams, William M. *Green Development: Environment and Sustainability in the Third World*. London: Routledge, 1990.

Altbach, Philip G. *The Knowledge Context: Comparative Perspectives on the Distribution of Knowledge*. Albany, N.Y.: State University of New York Press, 1987.

Anderson, John R. *Cognitive Psychology and Its Implications*. 2nd ed. New York: W. H. Freeman and Company, 1985.

Bakhtin, Mikhail M. *The Dialogic Imagination: Four Essays*. Edited by Michael Holquist. Translated by Caryl Emerson and Michael Holquist. Austin: University of Texas Press, 1981.

Basualdo, Isabel, Elsa Zardini, and Mirtha Ortiz. "Medicinal Plants of Paraguay: Underground Organs." *Economic Botany* 45 (1991): 86–96.

Baur, Susan. *The Dinosaur Man: Tales of Madness and Enchantment from the Back Ward*. New York: Edward Burlingame, 1991.

Bell, Robert. *Impure Science: Fraud, Compromise and Political Influence in Scientific Research*. New York: John Wiley and Sons, 1992.

Berlin, Brent, and Paul Kay. *Basic Color Terms: Their Universality and Evolution*. Berkeley: University of California Press, 1969.

Bickerton, Derek. *Language and Species*. Chicago: University of Chicago Press, 1990.

Block, Ned, ed. *Readings in Philosophy of Psychology*. 2 vols. Cambridge, Mass.: Harvard University Press, 1980–81.

Bohm, David, and F. David Peat. *Science, Order, and Creativity*. New York: Bantam Books, 1987.

Bolter, Jay David. *Writing Space: The Computer, Hypertext, and the History of Writing*. Hillsdale, N.J.: Lawrence Erlbaum Associates, 1991.

Brand, Stewart. *The Media Lab: Inventing the Future at MIT*. New York: Viking Press, 1987.

Briggs, John P., and F. David Peat. *Looking Glass Universe: The Emerging Science of Wholeness*. New York: Simon and Schuster, 1984.

Broad, William, and Nicholas Wade. *Betrayers of the Truth*. New York: Simon and Schuster, 1982.

Brokensha, David, D. M. Warren, and Oswald Werner, eds. *Indigenous Knowledge Systems and Development*. Washington, D.C.: University Press of America, 1980.

Brown, Judith. *Gandhi: Prisoner of Hope*. New Haven, Conn.: Yale University Press, 1989.

Bruner, Jerome. *Beyond the Information Given: Studies in the Psychology of Knowledge*. New York: W. W. Norton Company, 1973.

Buber, Martin. *I and Thou*. 2nd ed. Translated by Ronald Gregor Smith. New York: Charles Scribner's Sons, 1958.

Buchanan, Bruce G., and Edward H. Shortliffe, eds. *Rule-Based Expert Systems: The MYCIN Experiments of the Stanford Heuristic Programming Project*. Reading, Mass.: Addison-Wesley, 1984.

Carroll, Lewis. *Alice's Adventures in Wonderland; Through the Looking-Glass; The Hunting of the Snark*. New York: The Modern Library, n.d.

Cerf, Christopher, and Victor Navasky. *The Experts Speak: The Definitive Compendium of Authoritative Misinformation*. New York: Pantheon, 1984.

Churchland, Paul. *Matter and Consciousness: A Contemporary Introduction to the Philosophy of Mind*. Cambridge, Mass.: MIT Press, 1988.

____. "Perceptual Plasticity and Theoretical Neutrality." *Philosophy of Science* 55 (1988): 167–87.

Clark, Ramsey. *The Fire This Time: U.S. War Crimes in the Gulf*. New York: Thunder's Mouth Press, 1992.

Cohen, Robert, and Alexander W. Siegel, eds. *Context and Development*. Hillsdale, N.J.: Lawrence Erlbaum Associates, 1991.

Cole, Michael, and Barbara Means. *Comparative Studies of How People Think: An Introduction*. Cambridge, Mass.: Harvard University Press, 1981.

Collins, Harry M. *Artificial Experts: Social Knowledge and Intelligent Machines*. Cambridge, Mass.: MIT Press, 1990.

de la Court, Thijs. *Beyond Brundtland: Green Development in the 1990s*. Translated by Ed Bayens and Nigel Harle. New York: New Horizons Press, 1990.

Cowley, Geoffrey. "The Electronic Goddess." *Newsweek*, 6 March 1989: 50.

Cummins, Robert. "Inexplicit Information." *The Representation of Knowledge and Belief*. Edited by Myles Brand and Robert M. Harnish. Tucson: University of Arizona Press, 1986. 116–26.

Danforth, Kenneth C., ed. *Journey into China*. Washington, D.C.: National Geographic Society, 1982.

Davies, Graham M., and Donald M. Thomson, eds. *Memory in Context: Context in Memory*. New York: John Wiley & Sons, 1988.

Davis, Stephen D., Stephen J. M. Droop, Patrick Gregerson, Louise Henson, Christine J. Leon, Jane Lamlein Villa-Lobos, Hugh Synge and Jana Zantovska. *Plants in Danger: What Do We Know?* Cambridge, U.K.: International Union for Conservation of Nature and Natural Resources, 1986.

"The Day after Trinity." Produced and directed by Jon Else/KTEH, San Jose, Calif. John D. and Catherine T. MacArthur Foundation, 1980.

DeFrancis, John. *Visible Speech: The Diverse Oneness of Writing Systems.* Honolulu: University of Hawaii Press, 1989.

Dennett, Daniel. *The Intentional Stance.* Cambridge, Mass.: MIT Press, 1987.

Deregowski, J. B. "Real Space and Represented Space: Cross-cultural Perspectives." *Behavioral and Brain Sciences* 12 (1989): 51–119.

Dictionary of Computing. Edited by Edward L. Glaser, I. C. Pyle, and Vallerie Illingworth. Oxford: Oxford University Press, 1990.

Dörfler, Hans-Peter, and Gerhard Roselt. *The Dictionary of Healing Plants.* London: Blandford Press, 1989.

Dretske, Fred I. *Knowledge and the Flow of Information.* Cambridge, Mass.: MIT Press, 1981.

Eagle Walking Turtle. *Indian America: A Traveler's Companion.* 2nd ed. Santa Fe, Mexico: John Muir, 1991.

Edelman, Gerald M. *Bright Air, Brilliant Fire: On the Matter of the Mind.* New York: Basic Books, 1992.

Eiseley, Loren. *The Invisible Pyramid.* New York: Charles Scribner's Sons, 1970.

____. *The Night Country.* New York: Charles Scribner's Sons, 1971.

Eliot, T. S. "Tradition and the Individual Talent." *The Norton Anthology of English Literature.* Vol. 2. Edited by M. H. Abrams. New York: W. W. Norton, 1962. 1501–08. Originally published in 1919.

Erickson, Deborah. "Hacking the Genome." *Scientific American,* April 1992: 128–37.

Essien, Patrick P. "The Use of Annang Proverbs as Tools of Education in Nigeria." Ph.D. diss. St. Louis University, St. Louis, Mo., 1978.

Etzioni, Amitai. *The Moral Dimension: Toward a New Economics.* New York: The Free Press, 1988.

Evans, Jonathan St. B. T. *Bias in Human Reasoning: Causes and Consequences.* London: Lawrence Erlbaum Associates, 1989.

Facts on File: Weekly World News Digest with Index. New York: Facts on File, Inc.

Feigenbaum, Edward, Pamela McCorduck, and H. Penny Nii. *The Rise of the Expert Company: How Visionary Companies Are Using Artificial Intelligence to Achieve Higher Productivity and Profits.* New York: Times Books, 1988.

Fodor, Jerry A. "Observation Reconsidered." *Philosophy of Science* 51 (1984): 23–43.

____. "A Reply to Churchland's 'Perceptual Plasticity and Theoretical Neutrality.'" *Philosophy of Science* 55 (1988): 188–98.

Forje, John W. *Science and Technology in Africa.* Harlow, Essex, U.K.: Longman, 1989.

Foss, Donald, and David Hakes. *Psycholinguistics: An Introduction to the Psychology of Language.* New Jersey: Prentice-Hall, 1978.

Freeman, Walter J. "The Physiology of Perception." *Scientific American,* February 1991: 78–85.

Freire, Paulo. *Education for Critical Consciousness.* 1973. New York: The Continuum Publishing Company, 1987.

____. *Pedagogy of the Oppressed.* 1970. Translated by Myra Bergman Ramos. New York: The Continuum Publishing Company, 1988.

"Friendly Fire." Produced by Harry Radliffe. Correspondent Steve Kroft. "60 Minutes." CBS, New York. 10 November 1991.

"From the Heart of the World: The Elder Brothers' Warning." Produced and directed by Alan Ereira for BBC Productions, and edited for PBS *Nature* series, 1991.

(Unedited, original film is distributed by Mystic Fire Video, 225 Lafayette St., Suite 1206, New York, N.Y. 10012.)

Gadamer, Hans-Georg. *Truth and Method.* Translated and edited by Garrett Barden and John Cumming. New York: Seabury Press, 1975.

____. *Philosophical Hermeneutics.* Berkeley: University of California Press, 1976.

Gallant, Stephen I. "Connectionist Expert Systems." *Communications of the ACM* 31 (1988): 152–69.

Garrod, Simon, and Anthony Anderson. "Saying What You Mean in Dialogue: A Study in Conceptual and Semantic Coordination." *Cognition* 27 (1987): 181–218.

Geertz, Clifford. *The Interpretation of Cultures: Selected Essays.* New York: Basic Books, 1973.

____. *Local Knowledge: Further Essays in Interpretative Anthropology.* New York: Basic Books, 1983.

Gleick, James. *Chaos: Making a New Science.* New York: Viking Press, 1987.

Gluck, Mark A., and David E. Rumelhart, eds. *Neuroscience and Connectionist Theory.* Hillsdale, N.J.: Lawrence Erlbaum Associates, 1990.

Goody, Jack. *The Logic of Writing and the Organization of Society.* Cambridge: Cambridge University Press, 1986.

____. *The Interface between the Written and the Oral.* Cambridge: Cambridge University Press, 1987.

Gordon, Vance C., and Dale E. Wierenga, "The Drug Development and Approval Process." *New Drug Approvals in 1990.* Washington, D.C.: Pharmaceutical Manufacturers Association, January 1991.

Goulet, Denis. *The Cruel Choice: A New Concept in the Theory of Development.* New York: Atheneum, 1971.

____. *A New Moral Order: Studies in Development Ethics and Liberation Theology.* Maryknoll, N.Y.: Orbis Books, 1974.

____. *The Uncertain Promise: Value Conflicts in Technology Transfer.* 1977. New York: New Horizons Press, 1989.

Guillet, David, ed. *Expert Systems Applications in Anthropology, Part 1.* Spec. issue of *Anthropological Quarterly* 62 (1989): 57–102.

____. *Expert Systems Applications in Anthropology, Part 2.* Spec. issue of *Anthropological Quarterly* 62 (1989): 107–44.

Gutzwiller, M. C. "Quantum Chaos." *Scientific American*, January 1992: 78–84.

Hamill, James F. *Ethno-Logic: The Anthropology of Human Reasoning.* Urbana: University of Illinois Press, 1990.

Hartley, Karen. "Solar System Chaos." *Astronomy*, May 1990: 34–39.

Haugeland, John, ed. *Mind Design: Philosophy, Psychology, Artificial Intelligence.* 1981. Cambridge, Mass.: MIT Press, 1987.

Havelock, Eric A. *Preface to Plato.* Cambridge, Mass.: Belknap Press of Harvard University Press, 1963.

Hawking, Stephen W. *A Brief History of Time: From the Big Bang to Black Holes.* New York: Bantam Books, 1988.

Heider, Eleanor (Eleanor Rosch). "'Focal' Color Areas and the Development of Color Names." *Developmental Psychology* 4 (1971): 447–55.

Heim, Michael. *Electric Language: A Philosophical Study of Word Processing.* New Haven, Conn.: Yale University Press, 1987.

Hettne, Bjorn. *Development Theory and the Three Worlds.* New York: John Wiley & Sons, 1990.

Hopkins, Gerard Manley. *The Sermons and Devotional Writings of Gerard Manley Hopkins.* Edited by Christopher Devlin. London: Oxford University Press, 1959.

Hutchins, Edwin. *Culture and Inference: A Trobriand Case Study.* Cambridge, Mass.: Harvard University Press, 1980.

Huxley, Thomas H. *Essays upon Some Controverted Questions.* New York: Macmillan, 1892.

Illich, Ivan. "Outwitting the Developed Countries." *Underdevelopment and Development: The Third World Today. Selected Readings.* Edited by Henry Bernstein. Middlesex, U.K.: Penguin Books, 1973. 357–68.

Jackson, Peter. *Introduction to Expert Systems.* Wokingham, U.K.: Addison-Wesley Publishing Company, 1986.

Jackson, Peter, Han Reichgelt, and Frank van Harmelen, eds. *Logic-based Knowledge Representation.* Cambridge, Mass.: MIT Press, 1989.

Jacob, Margaret C. *The Cultural Meaning of the Scientific Revolution.* Philadelphia: Temple University Press, 1988.

Jamison, Ellen. *World Population Profile.* U.S. Department of Commerce: Bureau of the Census, 1989.

Jamison, Ellen, Peter D. Johnson, and Richard A. Engels. *World Population Profile.* U.S. Department of Commerce: Bureau of the Census, 1987.

John Paul II. *Redemptor Hominis* (The Redeemer of Man). Boston: The Daughters of St. Paul, 1979.

____. *Laborem Exercens* (On Human Work). Boston: The Daughters of St. Paul, 1981.

____. *Sollicitudo Rei Socialis* (On Social Concern). Boston: The Daughters of St. Paul, 1987.

____. *Centesimus Annus* (The Hundredth Year). Washington, D. C.: Origins, 1991.

Johnson-Laird, Philip N. *The Computer and the Mind: An Introduction to Cognitive Science.* Cambridge, Mass.: Harvard University Press, 1988.

Joshi, A. R., and J. M. Edington. "The Use of Medicinal Plants by Two Village Communities in the Central Development Region of Nepal." *Economic Botany* 44 (1990): 71–83.

Judson, Horace Freeland. *The Search for Solutions.* New York: Holt, Rinehart, and Winston, 1980.

Kandel, Abraham, ed. *Fuzzy Expert Systems.* Boca Raton, Fla.: CRC Press, 1991.

Kohn, Alexander. *False Prophets.* Oxford: Basil Blackwell, 1986.

Kosslyn, Stephen M. "Aspects of Cognitive Neuroscience of Mental Imagery." *Science* 240 (1988): 1621–26.

Kosslyn, Stephen M., and Olivier Koenig. *Wet Mind: The New Cognitive Neuroscience.* New York: The Free Press, 1992.

Kosslyn, Stephen M., and J. R. Pomerantz. "Imagery, Propositions, and the Form of Internal Representations." *Cognitive Psychology* 9 (1977): 52–76.

Kuhn, Thomas. *The Structure of Scientific Revolutions.* 1962. 2nd ed. Chicago: University of Chicago Press, 1970.

Lachman, Roy, Janet L. Lachman, and Earl C. Butterfield. *Cognitive Psychology and Information Processing: An Introduction.* Hillsdale, N.J.: Lawrence Erlbaum Associates, 1979.

Lakoff, George. *Women, Fire, and Dangerous Things: What Categories Reveal about the Mind.* Chicago: The University of Chicago Press, 1987.

Lappé, Frances Moore, and Joseph Collins. *Food First: Beyond the Myth of Scarcity.* Fourth Printing. New York: Ballantine Books, 1982.

Lebret, Louis-Joseph. *The Last Revolution: The Destiny of Over- and Under-developed Nations.* Translated by John Horgan. New York: Sheed and Ward, 1965.

Leith, Philip. *Formalism in AI and Computer Science.* New York: Ellis Horwood, 1990.

Lewis, David. *The Voyaging Stars.* Sydney: Collins, 1978.

Lewis, W. Arthur. *The Theory of Economic Growth.* London: George Allen and Unwin, 1955.

Lister, Richard G., and Herbert J. Weingartner. *Perspectives on Cognitive Neuroscience.* Oxford: Oxford University Press, 1991.

Lord, Albert B. *The Singer of Tales.* New York: Atheneum, 1968.

Luria, A. R. *The Man with a Shattered World: The History of a Brain Wound.* Translated by Lynn Solotaroff. New York: Basic Books, 1972.

____. *Cognitive Development: Its Cultural and Social Foundations.* Translated by Martin Lopez-Morillas and Lynn Solotaroff. Edited by Michael Cole. Cambridge, Mass.: Harvard University Press, 1976.

____. *Language and Cognition.* Edited by James Wertsch. New York: John Wiley and Sons, 1981.

McLuhan, Marshall. *The Gutenberg Galaxy: The Making of Typographic Man.* Toronto: University of Toronto Press, 1962.

McLuhan, Marshall, and Quentin Fiore. *The Medium Is the Massage.* New York: Bantam Books, 1967.

"MacNeil Lehrer News Hour." Public Broadcasting System. KETC, St. Louis, Mo., 1991.

McNeill, William H. *The Rise of the West: A History of the Human Community.* New York: Mentor, 1965.

Magill, Frank N., ed. *Masterpieces of World Philosophy.* New York: Harper Collins, 1990.

March, Kathryn S., and Rachelle L. Taqqu. *Women's Informal Associations in Developing Countries: Catalysts for Change?* Boulder, Colo.: Westview Press, 1986.

Medawar, Peter B. *The Threat and the Glory: Reflections on Science and Scientists.* Edited by David Pyke. New York: HarperCollins, 1990.

Minsky, Marvin. "A Framework for Representing Knowledge." *Mind Design.* Edited by John Haugeland. Cambridge, Mass.: MIT Press, 1987, 95–128.

Moore, Cristopher. "Unpredictability and Undecidability in Dynamic Systems." *Physical Review Letters* 64 (1990): 2354–57.

Morrish, Ivor. *Education Since 1800.* New York: Barnes and Noble, 1970.

Moser, Paul, and Arnold vander Nat, eds. *Human Knowledge: Classical and Contemporary Approaches.* New York: Oxford University Press, 1987.

Myers, Norman. *A Wealth of Wild Species: Storehouse for Human Welfare.* Boulder, Colo.: Westview Press, 1983.

National Research Council. Ad hoc panel. *Microcomputers and Their Applications For Developing Countries*. Boulder, Colo.: Westview Press, 1986.

___. *Cutting Edge Technologies and Microcomputer Applications for Developing Countries*. Boulder, Colo.: Westview Press, 1988.

National Security Strategy of the United States. Washington, D. C.: White House, March 1990.

"NBC Nightly News." NBC Network. KSDK, St. Louis, Mo., 1991.

Negoita, Constantin Virgil. *Expert Systems and Fuzzy Systems*. Menlo Park, Calif.: The Benjamin/Cummings Publishing Company, 1985.

Nozick, Robert. *Anarchy, State, and Utopia*. New York: Basic Books, 1974.

Ong, Walter J. *Ramus, Method, and the Decay of Dialogue: From the Art of Discourse to the Art of Reason*. 1958. Cambridge, Mass.: Harvard University Press, 1983.

___. *In the Human Grain: Further Explorations of Contemporary Culture*. New York: Macmillan, 1967.

___. *The Presence of the Word: Some Prolegomena for Cultural and Religious History*. New Haven, Conn.: Yale University Press, 1967.

___. *Orality and Literacy: The Technologizing of the Word*. London: Methuen, 1982.

Pacey, Arnold. *The Culture of Technology*. Cambridge, Mass.: MIT Press, 1983.

Pagels, Heinz R. *The Dreams of Reason: The Computer and the Rise of the Sciences of Complexity*. New York: Simon and Schuster, 1988.

Paul VI. *Populorum Progressio* (On the Development of Peoples). Boston: The Daughters of St. Paul, 1967.

Peat, F. David. *Superstrings and the Search for the Theory of Everything*. Chicago: Contemporary Books, 1988.

Penfield, Joyce. *Communicating with Quotes: The Igbo Case*. Westport, Conn.: Greenwood Press, 1983.

Pogge, Thomas W. "Rawls and Global Justice." *Canadian Journal of Philosophy* 18 (1988): 227–56.

___. *Realizing Rawls*. Ithaca: Cornell University Press, 1989.

Polanyi, Michael. *Personal Knowledge: Towards a Post-Critical Philosophy*. Chicago: University of Chicago Press, 1958.

___. *The Tacit Dimension*. Garden City, N.Y.: Doubleday, 1966.

Popper, Karl R. *The Logic of Scientific Discovery*. New York: Basic Books, 1959.

___. *Objective Knowledge: An Evolutionary Approach*. Oxford: Clarendon Press, 1972.

Pradervand, Pierre. *Listening to Africa: Developing Africa from the Grassroots*. New York: Praeger, 1989.

Proctor, Robert. *Value-Free Science?: Purity and Power in Modern Knowledge*. Cambridge, Mass.: Harvard University Press, 1991.

Quine, Willard Van Orman. *Word and Object*. New York: John Wiley and Sons, 1960.

___. *From a Logical Point of View: 9 Logico-Philosophical Essays*. New York: Harper Torchbooks, 1963.

Quine, Willard Van Orman, and J. S. Ullian. *The Web of Belief*. New York: Random House, 1970.

Rawls, John. *A Theory of Justice*. Cambridge, Mass.: Belknap Press of Harvard University Press, 1971.

____. "Kantian Constructivism in Moral Theory: The Dewey Lectures 1980." *The Journal of Philosophy* 77 (1980): 515–77.

____. "The Basic Liberties and Their Priority." *Liberty, Equality, and Law*. Edited by Sterling M. McMurrin. Salt Lake City: University of Utah Press, 1987, 1–88.

Rosch, Eleanor. "Cognitive Reference Points." *Cognitive Psychology* 7 (1975): 532–47.

____. "Linguistic Relativity." *Etc.* 44 (1987): 254–79.

Rosch, Eleanor, Francisco J. Varela, and Evan Thompson. *The Embodied Mind: Cognitive Science and Human Experience*. Cambridge, Mass.: MIT Press, 1991.

Rostow, W. W. *The Stages of Economic Growth: A Non-Communist Manifesto*. Cambridge: Cambridge University Press, 1960.

Rumelhart, David, and James McClelland. *Parallel Distributed Processing: Explorations in the Microstructure of Cognition*. 2 vols. Cambridge, Mass.: Harvard University Press, 1986.

Sacks, Oliver. *The Man Who Mistook His Wife for a Hat*. New York: Summit Books, 1985.

____. *Seeing Voices: A Journey into the World of the Deaf*. Berkeley: University of California Press, 1989.

St. Louis Post-Dispatch. St. Louis, Mo.: Pulitzer Publishing Company.

Schank, Roger C. *Tell Me a Story: A New Look at Real and Artificial Memory*. New York: Scribner, 1990.

Schumacher, E. F. *Small Is Beautiful: A Study of Economics as if People Mattered*. New York: Harper Colophon Books, 1973.

Scribner, Sylvia, and Michael Cole. *The Psychology of Literacy*. Cambridge, Mass.: Harvard University Press, 1981.

Searle, John R. "Indeterminacy, Empiricism, and the First Person." *The Journal of Philosophy* 84 (1987): 123–46.

Shweder, Richard A., and Robert A. LeVine, eds. *Culture Theory: Essays on Mind, Self, and Emotion*. Cambridge: Cambridge University Press, 1984.

Sifry, Micah L., and Christopher Cerf, eds. *The Gulf War Reader*. New York: Times Books, 1991.

Smith, Tony. "The Underdevelopment of Development Literature: The Case of Dependency Theory." *The State and Development in the Third World*. Edited by Atul Kohli. Princeton, N.J.: Princeton University Press, 1986. 25–66.

South Commission. *The Challenge to the South*. Oxford: Oxford University Press, 1990.

Soyinka, Wole. Keynote Address. 34th Annual Meeting of the African Studies Association. Adam's Mark Hotel, St. Louis, Mo. 23 November 1991.

Sreberny-Mohammadi, Annabelle. "Media Integration in the Third World: An Ongian Look at Iran." *Media, Consciousness, and Culture: Explorations of Walter Ong's Thought*. Edited by Bruce E. Gronbeck, Thomas J. Farrell, and Paul A. Soukup. Newbury Park, Calif.: Sage Publications, 1991. 133–46.

Sternberg, Robert J., ed. *Wisdom: Its Nature, Origins, and Development*. Cambridge: Cambridge University Press, 1990.

Stich, Stephen. *The Fragmentation of Reason: Preface to a Pragmatic Theory of Cognitive Evaluation*. Cambridge, Mass.: MIT Press, 1990.

———. *From Folk Psychology to Cognitive Science: The Case against Belief.* Cambridge, Mass.: MIT Press, 1983.

Stringer, Christopher B. "The Emergence of Modern Humans." *Scientific American*, December 1990: 98–104.

Swaby, Peter Alan. "VIDES: An Expert System for Visually Identifying Microfossils." *IEEE Expert* April 1992: 36–42.

Thomas, Lewis. *The Lives of a Cell: Notes of a Biology Watcher*. New York: Viking Press, 1974.

———. *The Medusa and the Snail: More Notes of a Biology Watcher*. New York: Bantam Books, 1979.

Tulving, Endel. *Elements of Episodic Memory*. Oxford: Oxford University Press, 1983.

Turner, Raymond. *Logics for Artificial Intelligence*. Chichester, U.K.: Ellis Horwood Limited, 1984.

Tversky, Amos, and Daniel Kahneman. "Judgment Under Uncertainty: Heuristics and Biases." *Science* 185 (1974): 1124–31.

Tyler, Stephen A. *The Said and the Unsaid: Mind, Meaning, and Culture*. New York: Academic Press, 1978.

———. *The Unspeakable: Discourse, Dialogue, and Rhetoric in the Postmodern World.* Madison: University of Wisconsin Press, 1987.

UNICEF. *The State of the World's Children*. Oxford: Oxford University Press, 1990.

Vygotsky, L. S. *Thought and Language*. Edited and translated by Eugenia Hanfmann and Gertrude Vakar. Cambridge, Mass.: MIT Press, 1962.

Wertsch, James. *Culture, Communication, and Cognition: Vygotskian Perspectives*. Cambridge: Cambridge University Press, 1985.

———. *Vygotsky and the Social Formation of Mind*. Cambridge, Mass.: Harvard University Press, 1985.

Whorf, Benjamin Lee. *Language, Thought, and Reality*. Cambridge, Mass.: MIT Press, 1956.

Williams, Rowan. *The Wound of Knowledge: Christian Spirituality from the New Testament to St. John of the Cross*. 2nd rev. ed. Cambridge, Mass.: Cowley Publications, 1991.

Wilson, Allan C., and Rebecca L. Cann. "The Recent African Genesis of Humans." *Scientific American* April 1992: 68–73.

Wilson, Edward O. *Biophilia: The Human Bond with Other Species*. Cambridge, Mass.: Harvard University Press, 1984.

Winner, Langdon. *The Whale and the Reactor: A Search for Limits in an Age of High Technology*. Chicago: University of Chicago Press, 1986.

Winograd, Terry, and Fernando Flores. *Understanding Computers and Cognition: A New Foundation for Design*. Reading, Mass.: Addison-Wesley Publishing Company, 1987.

Wittgenstein, Ludwig. *Philosophical Investigations*. Translated by G. E. M. Anscombe. Oxford: B. Blackwell, 1953.

Wolszczan, A., and D. A. Frail. "A Planetary System around the Millisecond Pulsar PSR1257+12." *Nature*, January 9, 1992: 145–47

World Commission on Environment and Development. *Our Common Future*. Oxford: Oxford University Press, 1987.

World Health Organization (WHO). *World Health Statistics Annual.* Geneva: World Health Organization, 1989.

Yates, Frances A. *The Art of Memory.* Chicago: University of Chicago Press, 1966.

Zadeh, Lofti A. "Fuzzy Logic." *Computer* April 1988: 83–93.

Ziman, John M. *Public Knowledge: An Essay Concerning the Social Dimension of Science.* Cambridge: Cambridge University Press, 1968.

____. *Reliable Knowledge: An Exploration of the Grounds for Belief in Science.* Cambridge: Cambridge University Press, 1978.

Zukav, Gary. *The Dancing Wu Li Masters: An Overview of the New Physics.* New York: Bantam Books, 1979.

Index

ABOUT THE AUTHOR

Doris M. Schoenhoff has an interdisciplinary Ph.D. focusing on Artificial Intelligence and international development, and has traveled worldwide as a computer software specialist.

ISBN 0-313-28821-6
90000>
EAN
9 780313 288210
HARDCOVER BAR CODE